Welcome To The Moon!

12 Lunar Expeditions
For Small Telescopes

by Robert Bruce Kelsey

Naturegraph Publishers

Library of Congress Cataloging-in-Publication Data

Kelsey, Robert Bruce, 1954-
  Welcome to the moon! : twelve lunar expeditions for small telescopes / Robert Bruce Kelsey.
    p.   cm.
  Includes bibliographical references and index.
  Summary: A guide to the moon for beginning astronomers with detailed information on all important sites as well as directions for using a small telescope with optimum viewing success.
  ISBN 0-87961-245-2 (alk. paper)
  1. Moon–Juvenile literature. 2. Moon–Observers' manuals –Juvenile literature.   [1. Moon. 2. Moon–Observers' manuals.]
  I. Title. QB582.K45   1997
  523.3–dc21   97-12303
                                                              CIP
                                                              AC

All illustrations and photos by Robert Bruce Kelsey

Naturegraph Publishers has been publishing books on natural history, Native Americans, and outdoor subjects since 1946. Please write for our free catalog.

Books for a better world

Naturegraph Publishers, Inc.
3543 Indian Creek Road
Happy Camp, CA 96039
(916) 493-5353

Table of Contents

To Caelin, Harry, Erin, Rob
—who may one day track in regolith
instead of mud.

Prologue

Time: About 4.5 billion years ago
Place: The young solar system

Dust clouds. Massive protoplanets. Huge asteroids. Eccentric orbits and close flybys. Gravitational chaos. And then....

A planetoid, possibly the size of Mars, wallops Earth with a glancing blow. Clumps of our young planet splatter out into space. Caught in Earth's gravity, they collide, compress, and a new object is formed. Our moon. The new satellite begins to cool. A crust forms. And then....

Meteoroids, huge and small, scour out impact basins and drill crater hole upon crater hole. The lunar crust shatters. Wave upon wave of lava spills out onto the pitted lunar surface. It fills the basins. It eats away crater walls and covers crater floors. Some craters are swallowed whole. Of others there's nothing left but a few small crags sticking out like the masts of ships lost beneath the churning lava waves....[1]

---

1. The planetary collision hypothesis described here, like the comparative ages of lunar features later in the book, is conjectural and subject to debate. As we discover more planets around distant suns and refine our theories of solar system formation, we may one day know the "truth" about the moon's origin. In the meantime, the collision hypothesis serves to remind us that the moon's history hasn't exactly been uneventful—as you'll soon see for yourself.

WELCOME TO THE MOON!

A planet wracked by cataclysmic collisions, inundated by colossal floods, and host to humankind's first landfall beyond our own water planet. Yet for many amateur astronomers and casual observers, the moon is just an annoying celestial object. For those who prefer nebulae and galaxies, the "deep-sky" objects as they are called, the moon is just massive, mobile light pollution. Big and bright, it blots out nebulae, star clusters, and galaxies for weeks at a time. The only thing good about it is it has phases where it isn't visible at all.

A rather impolite way to treat our closest neighbor in space. But when you're hunting galaxies no bigger than a dust speck on your eyepiece, even the slightest amount of stray light is enough to ruin an observing session. On the other hand, for those of us whose lives will remain meaningful even if we don't find a supernova in the Whirlpool Galaxy, the moon is probably the most entertaining object at our disposal.

The moon is the perfect object if you like to observe when the urge hits you. Many deep-sky objects can be easily seen only during specific times of the year. But the moon is conveniently positioned for observing at least two weeks every month. You don't have to drive for hours to find dark skies, because light pollution isn't as bothersome when viewing the moon. You don't have to wait for a specific season or stay up half the night waiting for your target constellation to rise. Unless you want to wait up for a late last quarter feature, you can get in some fine observing and still keep your day job.

And the view is always changing. As the earth revolves around the sun and the moon circles the earth,

| Direction of Earth's Orbit Around Sun | Direction of Sun's Rays |

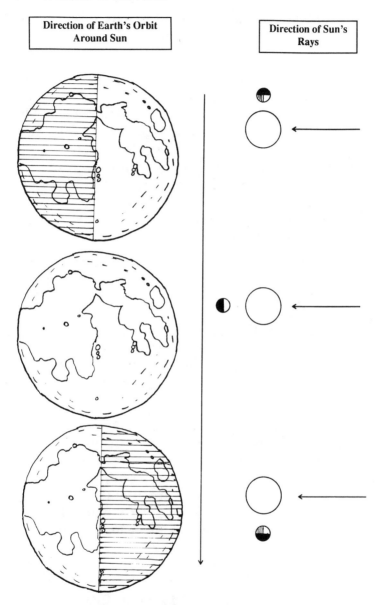

Fig. 1. The phases of the moon: First quarter, full, and last quarter.

the moon passes through phases. In each phase—new, first quarter, full, and last quarter—different segments of the lunar surface are visible. In the first quarter, the east side of the moon is lit up; in the last quarter, the west limb is illuminated; and at full moon the sunlight falls on the entire lunar surface. Since many calendars list moon phases, planning out your observing nights is simple and inexpensive.

But it's not just how much lunar terrain you can see that makes moon watching interesting. The angle of the sunlight changes throughout the lunar cycle and reveals different features of that terrain.

As the moon waxes from new to full, the "sunrise" (also known as the "terminator") moves across the lunar surface from lunar east to lunar west. A crater on the moon is illuminated first from the east, then from straight on. As the full moon wanes through the end of the last quarter, the "sunset" moves westward, and that crater shows its western profile to observers.

Catch an open star cluster once under excellent skies, and you've seen just about all there is to see. Look at a lunar feature one night only, and you've missed at least half of the details.

This book will take you on 12 lunar expeditions. They can all be completed as the moon waxes from new to full. You'll explore the lunar seas and the most prominent craters. You'll scan most of the Apollo landing sites. You'll even do a little lunar archeology on the way. In many expeditions, you'll retrace your steps from the previous trip. That way you can watch for changes in the scenery on your way to explore new territory.

Every expedition in this book is suitable for a small telescope. "Small" means a 2.4 inch (or 60mm) refractor up to a 4 inch reflector. All of the observation drawings were made with 60mm and 80mm refractors and a 4.25 inch "rich field" reflector, using plössl eyepieces and in some cases a 2x barlow lens. So, if you have an inexpensive department store refractor with reasonably good eyepieces, you can still see everything in this book. Or if you own a 4 inch equatorial reflector with clock drive, filters, and a 7-element, 6 millimeter eyepiece that cost more than your monthly car payment, you'll find it easier to see all the details.

Before you start off on your first lunar landing, you need to be honest with yourself: are you an experienced observer or not? Do you know how to align your finder scope? Do you know what averted vision is? Do you know about magnification and seeing conditions? If you do, skip ahead to the section *Learn the Landmarks Before You Cast Off*. But if you haven't done much observing, *For the New Navigator* will give you some tips to help make your first launch successful.

Chapter 1
For the New Navigator

If you are brave, clever, and not easily frustrated, you can skip this chapter. Just wait for nightfall, plunk your telescope down in your backyard, and start observing. What you'll see, well, that's another matter entirely.

After all, you'll be looking through a hole 5, 10, maybe 20 millimeters wide, at an object more than 221,000 miles away. And what you see through those key-hole sized eyepieces is moving—it can drift out of your field of view in the time it takes you to change from one eyepiece to another. If that weren't challenge enough, the world at the end of your telescope tube may be upside-down or reversed left to right!

So don't expect your first viewing sessions to be like watching your 21-inch television, popcorn in one hand and remote control in the other. Instead, think of it as an expedition in search of treasure. You sail across miles of space to get to that island of reflected sunlight above you. Along the way, you'll have to make course corrections as the moon appears to drift out of your field of view. You know that there are treasures hidden somewhere in the glare of the lunar surface. But you'll need good observing skills to find them.

You'll have to learn to steer, and you'll have to learn how to see things through that little hole at the end of your telescope. The first step is learning how to think in a world where everything is upside-down, or sideways, or both.

If you can read this without turning the book upside down you're already half way there!

Fig. 2. Upside-down reverse image.

The View Through the Finder Scope

Most telescopes in the 2.4 inch (or 60 millimeter) to 4 inch range come with a straight-through finder scope. This small telescope usually provides 5x to 7x magnification and shows you a wide field of view. It also presents the image upside-down.

The finder scope is commonly used for deep-sky observing, where you need to locate dim guide stars before looking for even dimmer objects in the main telescope. But the finder is also a convenience for lunar observing. If your main telescope is badly out of focus, you can easily mistake lunar glare for the lunar disk, and waste time trying to focus on thin air.

If you look at a telephone pole through the finder, it looks like figure 3.

If you looked at that same telephone pole through the main telescope, it would look right side up. In a refractor with a star diagonal, the image would be reversed left-to-right. In most reflectors, it would appear correct, but tilted at an angle based on where the focuser was on the telescope tube.

The telephone pole looks bigger in the main telescope because the main telescope magnifies objects more than the finder scope does. But when you magnify an object, the field of view shrinks. That's why

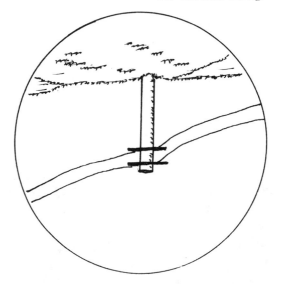

Fig. 3. A telephone pole viewed through a finder scope.

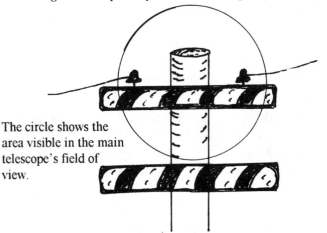

The circle shows the area visible in the main telescope's field of view.

Fig. 4. The same telephone pole viewed through a refractor's star diagonal.

you can see only one of the cross bars in this example. The other one is out of the field of view. So as you switch between the finder and the main telescope, you have to switch from thinking in an upside-down world to one where things are right-side-up, but may be swapped left to right. As a warm-up for this mental gymnastics, you should align your finder scope. Here's how:

Step 1: Pick an object far off in the distance to help you align the scopes. The top of a telephone pole or the peak of a roof is perfect.

Step 2: Find that object in your main telescope, and get it right in the center of the field of view.

Step 3: Now look through the finder scope. If the object in the main telescope is in the center of the finder scope's field, the two scopes are aligned. You don't have to do anything more here, so skip ahead to *Practice Steering Your Telescope.*

Step 4: If the object is not in the center of the finder's field of view, loosen all the screws that hold the finder scope in its circular brace.

Step 5: Move the finder scope until the object falls in the middle of the finder's field of view.

Step 6: Hold the finder scope tightly in place with one hand. With your other hand, tighten all the screws that hold the finder scope in its circular brace.

Step 7: Now check the view in the main telescope again. You might have jiggled the telescope while you were adjusting the finder. If the views in the main telescope and finder scope don't match, do Steps 2 through 7 until you get the finder aligned.

Practice Steering Your Telescope

With the finder aligned properly, it's time to learn how to steer. Your telescope may have one slow motion control or two controls. It may be an alt-azimuth mount, which moves the telescope up and down, right to left. It may be an equatorial mount, which lets you move the telescope in the same directions as stars move. If your telescope came with instructions on how to use the mount, read them over before you start. If you don't have any instructions, don't worry! You learn best by doing, not by reading.

So while it's daylight, take out your telescope. Start by turning the motion control knobs. Watch how the telescope moves. Which knob moves the telescope up and down? Which one moves it right to left? What happens when you turn the motion controls to the left? To the right?

Now that you know what the controls can do, find an object off in the distance, like the telephone pole you used to align the finder scope. Find it in the finder scope, then in the main telescope. Start at the top of the object, then use the motion controls to scan all the way down the object. Then start at the left side of the object, and scan all the way to the right. (This one is tricky. Remember the image in your main telescope may be swapped right to left!)

It takes a while to master the motion controls. Don't get discouraged. Keep at it until you can "steer" the main telescope in any direction you want to. The more you practice now, in the daytime, the easier it will be at night. And you'll spend your time observing instead of fumbling with knobs. Hundreds of small telescopes have been banished to sheds and garages because their owners couldn't "find anything" in their first observing sessions. Well, just whose fault is that?

Looking Through a Telescope

In the last section you learned how to find an object in your telescope and how to keep it in sight. Here are some tips on how to make that object give up all its hidden secrets.

Start with Low Magnification

Always use your lowest power eyepiece when you start looking for an object. High magnification narrows your field of view, so it is harder to get your bearings.

Eyepieces are measured in millimeters: 6mm, 12mm, 26mm, etc. The larger the eyepiece, the lower the magnification. To figure out the magnification of any eyepiece, divide the focal length of your telescope by the focal length of the eyepiece. (You may find the focal lengths listed on the box your telescope came in, or in the assembly instructions, or in the sales material you received when you purchased the telescope.)

Magnification = focal length of main telescope in millimeters ÷ focal length of eyepiece in millimeters.

Suppose you have a 60mm refractor with a focal length of 700. You have two eyepieces, 20mm and 6mm. The 20mm eyepiece gives a magnification of 35x (or 700 divided by 20). The 6mm eyepiece provides a 116x magnification. The highest useful magnification is usually about 2.5 times the size of your objective in millimeters. Here, that's 2.5 multiplied by 60, or 150. The highest useful magnification is useful only under perfectly clear, calm skies. You'll probably be happier with that 20mm eyepiece under most sky conditions.

Match Your Magnification to the Seeing Conditions

Wind and upper atmosphere conditions will affect how much you can see in your telescope. Looking through a telescope under windy conditions is at best annoying. At worst, it calls for a dose of motion sickness medication. If you can set up your telescope behind a wind barrier, you can usually get in some good medium power observing of the moon. Avoid high magnification on windy nights—that 73x eyepiece may make Mare Crisium look closer, but it also makes the slightest wobble seem like a moonquake! Atmospheric steadiness is important, too. If the stars are twinkling wildly when you first look up, the upper atmosphere is turbulent. This will cause light to scatter on its way through the atmosphere. Under high magnification, what should be a finely etched crater rim will look more like a brightly lit but very soggy pasta noodle. Stick to low magnification, where the effect of turbulence isn't quite as irritating.

Sky transparency affects how much light your telescope can collect. Dust, dew, clouds, and light pollution will block out or wash out the light reflecting off lunar features. You can see lunar features in just about any sky conditions, but naturally the darker and clearer your skies, the better.

Let Your Telescope Do the Work

Look through the eyepiece as if you are just looking across the room. Don't squint, don't stare, and don't hold your breath. Stay relaxed. Squinting and staring tire your eye quickly; a tired eye can't see as well. And don't forget to breathe! If you forget to breathe, you'll tense up. You don't see as well when you are tense.

Let Your Eyes Dark Adapt

Don't expect to go from a brightly lit room directly to the eyepiece and see tiny details on the lunar surface. Your eyes take 15 to 30 minutes to adjust to "night vision." You don't need fully dark adapted eyes to view the moon, but give your eyes a few minutes to recover from those 300 watt halogen floor lamps before you start looking for shadows by crater rims.

Use Both Eyes

You can only use one eye at a time when looking through a telescope. But you should switch back and forth between your eyes. This will keep you from getting tired as quickly.

Look Straight On for Some Objects, To the Side for Others

If you are looking at a crater on the moon, you should look directly at the object. Your eyes pick up coarse contrasts and colors better if you look straight at the object. But if you are trying to catch sight of a dim or tiny detail—say, the edge of a mountain range under glancing lighting—look just to the side of the object. This is called "averted vision" and it takes some getting used to. But you can practice anytime. Find a street sign or a billboard and look at one letter in the middle of one word. Now try to "see" the other letters and words without looking directly at them. The more you practice "seeing to the side" the better your averted vision will be. And the more you'll see in the telescope.

Don't Be Afraid to Adjust the Focus

Remember everything you see in your telescope is three-dimensional. Details that stand out at one focus may hide other details that will appear if you change the focus just slightly. Always start with a crisp focus of a known object. Before you scan along the lunar surface, first find a bright crater or a lunar seashore and bring that into focus. Now as you scan, you'll be able to see the "surface" details clearly.

Once you locate the object, try adjusting the focus slightly. Details below a crater rim may appear only if you take the rim just slightly out of focus.

Draw What You See

If you just take a "quick look" at an object, you'll probably miss it. Even the smallest lunar crater hides fascinating details, if you're willing to look for them.

The first time you catch an object in your telescope's field of view, it's like walking into a room for the first time. The first thing you notice is the big, obvious stuff—the sofa, a few chairs, the bookcase. If you walk out of the room right away, you probably won't remember anything more than that. You won't have had time to see the hazy light coming through the window curtains, or the color or patterns of the fabric on the sofa.

But if you stay longer in the room, your eyes will get adjusted to the room's "floor plan" and you'll start to see beyond the obvious to the small and detailed. Drawing an object helps you get past the sofa and the bright lamp in the corner so you can see the patterns and colors "hiding" around them. Anyone can see a crater hole by a mountain, just as anyone can see a sofa by a bookcase. But what's on the shelves? Are there crater holes in that mountain rising up near the crater? And where, exactly, are those crater holes?

Drawing what you see also gives you a record of what an object looked like at a specific time. This is important for lunar observing, where the angle of the sunlight can make an object seem to disappear from one night to the next.

To help you draw objects at the eyepiece, you'll need a low brightness flashlight (preferably with a red bulb in it), heavy weight drawing paper, and a clipboard. Many stores and catalogs that carry telescopes also sell red incandescent or LED observer's lights. But a little red enamel paint on a pen light will do just fine. There's a moon observation report form later in this book (see page 112), which you can reproduce in quantity on any copier machine. Most copier machines use a polished 20 pound weight paper which is fine for your first drawings. Eventually you may want to move up to artists sketching paper or less expensive, 24 pound ink-jet printer paper.

A Word About Filters

It is irónic that you need dark adapted eyes to see well and yet the brightness of the moon can overwhelm dark adapted eyes. Some people use filters to block some of the light reflected off the lunar surface. Photographic equipment retailers who sell telescopes as well as most mail order telescope retailers sell a variety of filters. Some are designed to dampen all wavelengths, and simply reduce the overall brightness of the lunar surface. Others are selective and are often credited with helping bring out details on the surface. Filters are a matter of personal preference. Blue and yellow filters are the traditional choices for lunar observation, but many observers prefer orange, yellow-green, and green. (Orange and yellow-green filters were used for the high magnification drawings in this book.)

If you already own a set of filters, by all means experiment with them as you work through this book. If you do not own filters but would like to try one or two, first make sure you have the right equipment! Not all small telescope eyepieces are threaded to accept filters. Most 1.25 inch eyepieces are threaded. Some of the .965 inch eyepieces, especially the ones that come with inexpensive refractors, are not.

Chapter 2
Learn the Landmarks Before You Cast Off

In the expeditions later in this book you'll cross the major lunar maria, traverse the larger mountain ranges, and peer down into the most prominent craters. Take a few minutes now to familiarize yourself with the names and locations of some of the stops on these lunar tours.

The Lunar Maria

The lunar seas, or maria (one sea is a "mare"), are the most obvious lunar features. Visible to the unaided eye, they're the eyes and mouth of the Man in the Moon. (Some people see them as the ears and body of a rabbit, or the bodies of Jack and Jill.) Despite their bluish color, they aren't water seas—they're lava flows. About 4.5 billion years ago, according to one theory, a small planetoid grazed the earth, spewing chunks of earth into space. Caught in the earth's gravity, these terrestrial shavings eventually reformed as our satellite.

For millions of years, meteoroids of all sizes bombarded the moon. Some meteoroids scooped out the impact basins we now call seas. Subsequent volcanic activity and new meteoroid impacts caused successive lava eruptions across the surface. The lava congealed in the impact basins, leaving us the seas we see today. In the lunar expeditions that follow, you'll take a look at these seas and some of the craters they flooded.

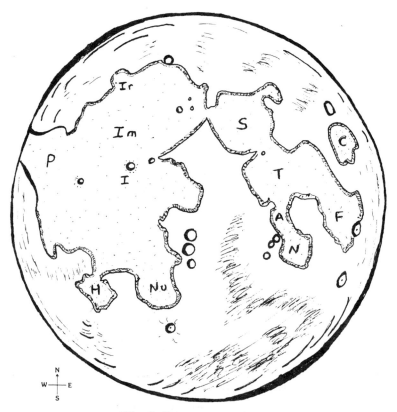

Fig. 5. Lunar seas and bays.

*The prominent maria on the lunar surface are listed in this normal view drawing of the full lunar surface. North is to the top, east is at the right. Mare Crisium (**C**) is in the lunar northeast. Below it lie Mare Fecunditatis (**F**) and Mare Nectaris (**N**). Directly west of Crisium is Mare Tranquillitatis (**T**) and above Tranquillitatis lies Mare Serenitatis (**S**). The channel where Tranquillitatis flows into*

*Nectaris is known as Sinus Asperitatis (**A**). Oceanus Procellarum (**P**) lies in the far lunar west, with Mare Insularum (**I**) to its east. Mare Imbrium (**Im**) stretches north from Insularum and merges with Sinus Iridum (**Ir**), actually a flooded crater basin, in the north-northwest. South of Insularum lie Mare Humorum (**H**) and Mare Nubium (**Nu**).*

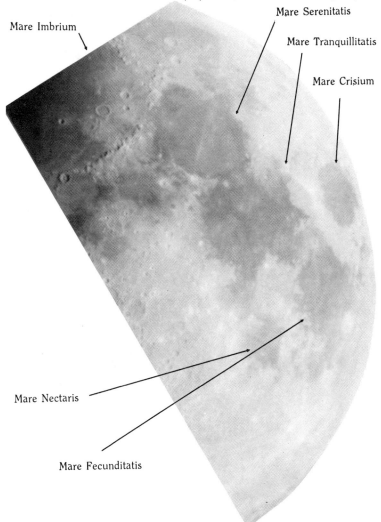

Mare Serenitatis

Mare Imbrium

Mare Tranquillitatis

Mare Crisium

Mare Nectaris

Mare Fecunditatis

Photo 1. The eastern maria.

The Lunar Mountain Ranges

The mountain ranges on the moon are not like the earth's mountain ranges. On Earth, mountain ranges are pushed up into the air when plates in the earth's crust collide. On the moon, the mountains are the rims of huge, ancient impact craters. Some mountain ranges have been damaged by impacts that came after they were formed. In the lunar expeditions that follow, you'll learn how to tell the history of a mountain range by the way it looks.

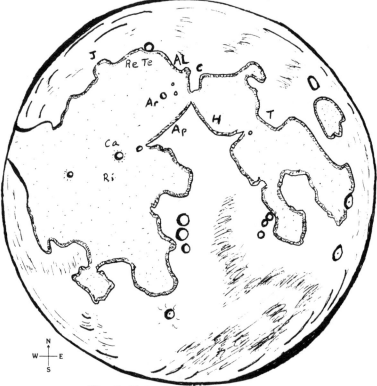

Fig. 6. Major lunar mountain ranges.

Some of the prominent mountain ranges on the lunar surface are listed in this normal view drawing of the full lunar surface. North is to the top, east is at the right. The Montes Jura (J) in the northwest are actually the rim of

Sinus Iridum. East of the Jura are the Montes Alpes (Al) and Montes Caucasus (C) in the north and Montes Archimedes (Ar), Montes Apenninus (Ap), and Montes Haemus (H) in the south. Between Mare Crisium and Mare Serenitatis lie Montes Taurus (T). In the western maria lie four small ranges, difficult to find in a small telescope but important pieces in the puzzle of the geologic history of the moon: Montes Teneriffe (Te), Montes Recti (Re), Montes Carpatus (Ca), and Montes Riphaeus (Ri).

The Major Lunar Craters

Much of the early history of the moon lies buried under the lava flows that filled the seas and many craters. But in some areas, clues about the ancient moon are still visible: old, flooded crater holes and the remains of mountain peaks near new craters. The major craters on the moon are easy to study even in a small telescope. Some are old, some are young, but all give you the chance to look back in time at the formation of our satellite. You'll visit all the craters shown in figure 7 later in this book.

When East Isn't, And Moving Left Takes You Right

Joe is out with his new 80mm refractor observing the crescent moon. He sees an interesting feature at the left edge of his field of view. He nudges the telescope tube to the left. Suddenly he's looking out into interstellar space, no moon in sight. What happened?

He forgot to compensate for the refractor's image reversal, of course. If you have practiced steering your telescope, you already know how your telescope presents images. The finder shows an upside-down image. A refractor with a star diagonal shows an object right-side-up, but image reversed. Reflectors show a tilted.

but correct image. You won't make Joe's mistake, right?

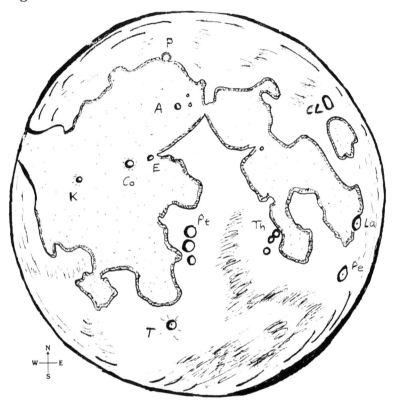

Fig. 7. Lunar craters.

Some of the prominent craters on the lunar surface are listed in this normal view drawing of the full lunar surface. North is to the top, east is at the right. In the east lie Cleomedes (**Cl**), Langrenus (**La**), and Petavius (**Pe**). West of Langrenus is the Theophilus crater chain (**Th**), and farther west still lies the Ptolemaeus crater chain (**Pt**). Tycho (**T**) is visible south of Ptolemaeus. In the west central maria lie Kepler (**K**), Copernicus (**Co**), and Eratosthenes (**E**). Above them at the eastern end of Mare Imbrium lies the Archimedes trio (**A**). The northernmost crater is Plato (**P**).

Just to make sure, you'd better break that typical earthling habit of thinking north is someplace beyond Greenland and west is where the sun sets. Lunar directions are based on the perspective of someone standing on the lunar surface. The moon always shows the same face to the earth as it orbits our planet, so the landmarks don't rotate or disappear completely from view. Mare Crisium lies in lunar east, no matter which way the crescent moon is pointing. Mare Serenitatis is lunar east of Mare Imbrium, even though on the full moon Imbrium looks more easterly.

The diagrams of the seas, mountains, and craters you just looked at show the moon as it appears to the unaided eye: the moon looks right-side-up, and lunar west is to the left. The full-moon maps of the expedition targets later in this book also take the "normal" perspective. But many of the drawings in this book show the lunar surface as it appears in a refractor with a star diagonal: lunar east is to the left, lunar north is to the top. Reflector owners will see the same images, but reversed back to "normal." Other drawings show a lunar feature as it appears through a reflector, so refractor owners will see the same image, but reversed.

Had enough pre-flight preparation? Fine. It's launch time.

Chapter 3
Lunar Expeditions:
New Moon Through First Quarter

Mare Crisium is easy to find in the first half of the lunar cycle. You'll use it as your launch pad for the first two expeditions, which take you on a tour of some of the prominent craters in the eastern portion of the moon. Then you'll visit two Apollo landing sites. And finally you'll come full circle and end the new moon expeditions with a detailed look at Crisium.

Star observers use "star hopping" to move from star to star on their way to interesting objects. Lunar observers "crater hop" instead. In this first expedition, you'll learn to crater hop while you explore one of the more diverse crater chains on the moon.

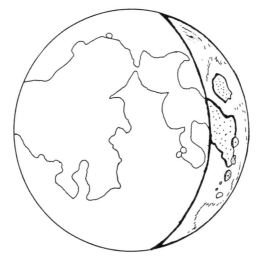

Expedition 1:
Hop a Crater Chain on the Three-Day-Old Moon

What to Look For: From Crisium, look south for a line of large craters marked as the Langrenus and Petavius regions on figure 8. The crater Langrenus is the large oblong crater on the east edge of Mare Fecunditatis. Unlike many other craters we'll see, its walls are still intact. Look for differences in brightness along the inside of the crater rim. Under the right light conditions and high magnification, you can see that these rings are actually terraces in the inner crater wall. High magnification will also show you the jagged edges of the crater rim and the two central peaks in the crater's center.

Vendelinus, south of Langrenus, is older and has suffered some battering by space debris. The crater walls don't show up as well as those of Langrenus. Under low power, it looks smoother than Langrenus—almost like it had been filled in. Actually, it has been. Around the outer rims of Vendelinus you will see several smaller crater holes. Lohse, Lamé, and Holden are the easiest to identify. The impacts that created

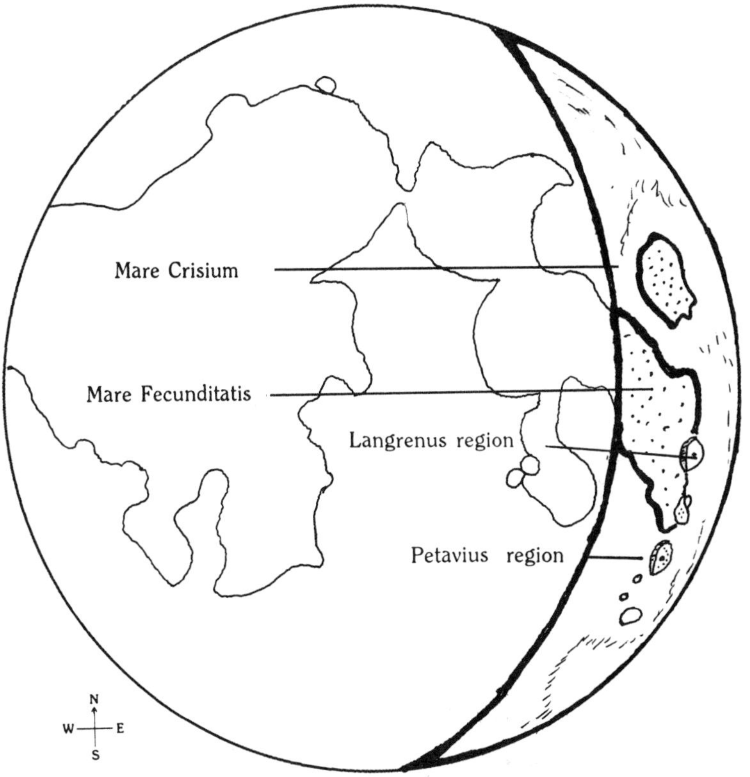

Fig. 8. Three-day crescent moon showing the Langrenus and Petavius regions.

these and other nearby craters spewed debris into the Vendelinus crater.

Seeing Lunar History

How old are these craters? We can't know for sure without getting rock samples. But observational evidence can help you determine relative crater age. "Younger" craters usually have well defined walls and rims. "Older" craters are smoother around the edges and are often shallow. Compare the crater Holden with the crater Lamé. Which looks like the newer crater?

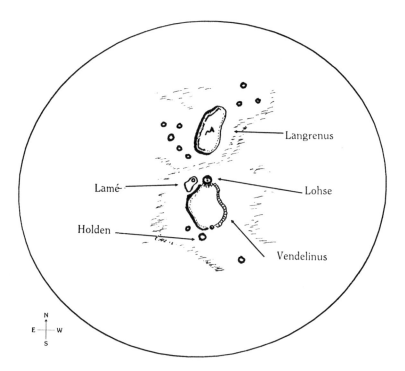

Fig. 9. The Langrenus region, 60mm refractor @ 82x.

South of Vendelinus lies Petavius. At 177 kilometers across, it is the largest of the three craters in this region. Its central peak is also larger than the one in Langrenus. Petavius's walls have been damaged by other impacts, and under some light angles you can see rilles in its floor, indicating lava flows. (Disagreement exists about whether this lava is mare lava from Fecunditatis or whether it was extruded from within Petavius itself.) Its neighbor Wrottesley, to the northwest, smashed into Mare Fecunditatis so close that its wall merged with the rim of Petavius. You can easily see its central peak.

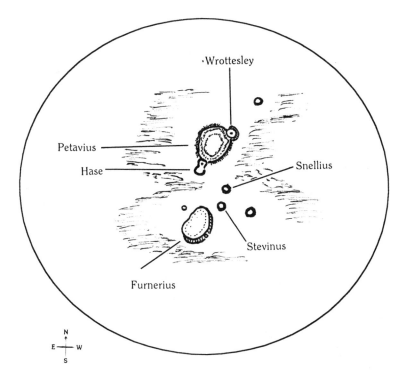

Fig. 10. The Petavius region, 60mm refractor @ 82x.

Seeing Lunar History

Large, shallow craters like Petavius are sometimes called "ring mountains" in early studies of the
moon. It is a reasonable interpretation, as there is no
round depression like you'd expect to see in an impact crater. All that is left are stubby ridges sticking
up out of relatively flat terrain. But both Wrottesley
and Petavius are impact craters. Those stubby ridges
around Petavius tell the story of a very old crater.

Mare Fecunditatis is one of the oldest impact basins on the moon, more than 3.9 billion years old.
Lunar geologists believe this and other impact basins
were filled with lava over the next 900 million years
or so. Looking through your telescope some 3 billion

years later, you can see that the old Petavius crater hole looks like it has been filled in with Fecunditatis lava. It's a good guess that the smaller object that created Petavius hit the moon after a much larger object had scooped out the Fecunditatis basin, but before the basin filled with lava. Whatever drilled out Wrottesley came much later.

Not all old craters are filled with lava. Some are battered almost beyond recognition. Hase, to the south of Petavius, is a good example of a disintegrated crater. What looks like a central peak in Hase is actually the wall of the crater Hase A inside Hase itself.

Snellius and its brighter companion Stevinus are the guideposts to the fourth large crater in this series, Furnerius. They are relatively new craters in the area called the Vallis Snellius. (This "valley" is a line of decayed craters that were created when a meteoroid scoured out the Mare Nectaris basin. Chunks of the basin were scattered south and east of Mare Nectaris, creating the jagged terrain around these craters.) Snellius does not have a central peak; Stevinus does. The "peak" in Snellius is really the rims of several old craters inside Snellius.

The crater Furnerius marks the end of this crater chain on the new moon. Furnerius is 125 kilometers across, a little smaller than Langrenus. Its rim is badly decayed—look for the numerous small crater holes on the rim, and for Furnerius B, the small crater inside Furnerius itself.

Notice anything peculiar about Furnerius compared to Petavius? Both are shallow, wide craters. The shallow plain inside Petavius can be explained by lava flows from Fecunditatis. But Furnerius is far away from that mare's shore. So why is its plain so shallow?

If you think there's another crater-aging process at work, you're right. When an object hits the lunar surface, the lunar crust is pulverized and thrown out around the impact hole. As more and more craters form around an older crater, more and more debris is deposited along the old crater's walls and floor. Sometimes the walls are destroyed, as they were in Hase's case. Sometimes over the ages the crater is slowly buried under not lava but "moon soil" scattered by other impacts. In the next expedition, you'll see how this ejected lunar material, called ejecta, can be used to comparatively date craters.

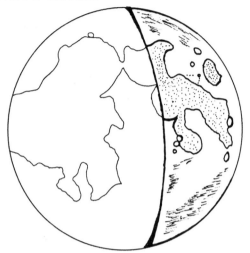

Expedition 2:
Blankets and Rays on the Five-Day-Old Moon

As the moon waxes from new to first quarter, the angle of the sunlight increases. Some features that were invisible in the low sun angle of the three-day moon, now are visible. And the terminator has moved farther west across the moon's surface, exposing more of Mare Tranquillitatis and Mare Nectaris.

What to Look For: Start this expedition at Mare Crisium. Later on you'll explore this region more carefully, but right now look inland from the western Crisium shoreline and find the crater Proclus. It can be difficult to spot on the three-day-old moon, but on the five-day-old moon it's easy to find. And so are the two rays, fanning out toward the eastern rim of Mare Tranquillitatis. The Proclus ray system is debris spewed out of the crater by the impact. This is also another clue to the relative age of craters. You'll see another example of this "ejecta" when you reach Theophilus.

From Crisium, drop south to the crater Langrenus, and then scan west beyond Mare Fecunditatis to the western shores of Mare Nectaris. There you will see a trio of craters: Theophilus, Cyrillus, and Catharina. These are the showpieces of the mid-new moon. And they're large enough to show quite a lot of detail even in a small telescope.

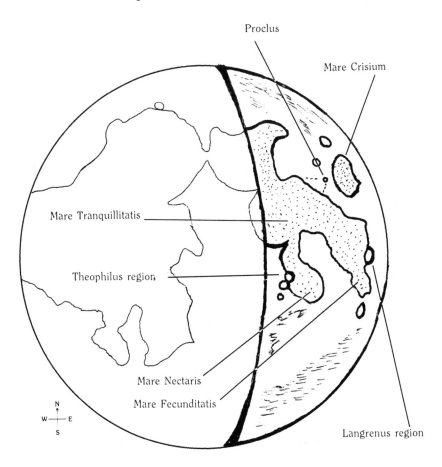

Fig. 11. Five-day crescent moon revealing the Theophilus region.

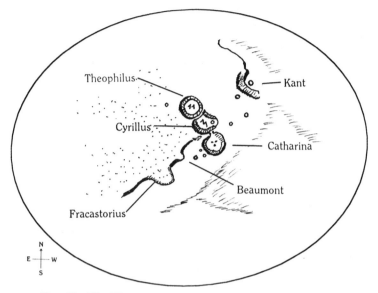

Fig. 12. The Theophilus region, 60mm refractor @ 82x.

Theophilus is the youngest of the three. It has three central peaks, one of which is small and lies hidden behind the two taller peaks. Its walls are well defined, and higher magnifications will show you these walls are terraced inside the crater. The outer edge of the crater rim is very bright. This is the ejecta blanket, in effect a layer of "moon soil" that was excavated out of the crater hole by the force of the impact.

Seeing Lunar History

The lunar surface is covered by a layer of rocky debris called regolith. This lunar soil is made up of dust from impacts, grains and pebbles sheared off larger rocks by tiny meteorite impacts, and droplets of molten rock and shards of lunar bedrock hurled out of impact craters. Beneath this regolith lies either mare lava or layers of impact debris depending on the location. Beneath this layer lies the lunar bedrock.

When a meteoroid slams into the crust, the impact scatters the existing regolith in all directions, gouges out chunks from the deeper layers of the lunar surface, and melts some of the material beneath the impact point. Astronauts Shepard and Mitchell found such "chunks" scattered all around the rim of Cone Crater during their Apollo 14 mission. Neil Armstrong saw what looked like droplets of melted rock inside small raised rim craters at the Apollo 11 landing site. If the object hitting the moon is very large, the force of impact can scatter debris far away from the actual impact. (Just how far will become apparent later on when you look at the rays on the full moon.)

Some of the material excavated by an impact falls like a blanket around the impact point. The bright rim of Theophilus is a perfect example. Larger chunks of impact debris can be hurled out of the impact zone with enough force to cause secondary cratering when they impact on the surface. Some of the material scattered by an impact falls linearly along the lunar surface, like the rays you just saw near Proclus. (On a later expedition, you'll see a different sort of ray pattern around the crater Aristillus.) Older lunar material collects near the crater hole; younger material falls farther away. What you see as Theophilus's ejecta blanket is really the lunar substratum that existed millions of years before some meteoroid gouged out the Theophilus crater.

The meteoroid that created Theophilus landed right next to a much older crater called Cyrillus. Whatever ejecta blanket Cyrillus used to have has been buried by regolith and the Theophilus impact debris. Cyrillus's northeast wall was damaged by the Theophilus impact. Southwest of the two visible central peaks is

Cyrillus A, a bright little crater hole from a much more recent impact. Like the crater Petavius, Cyrillus is often called a ring mountain because its rim is so badly damaged and its floor looks level with the surrounding terrain.

South of Cyrillus is what little is left of Catharina, another "ring mountain." Its inner rim appears wider than the rims of its neighbors, partly because the rim has itself been flattened out by subsequent impacts. You can see the results of these impacts. There's almost no rim visible on the north and south ends of Catharina. Based on what you saw in the Langrenus and Petavius regions, which is the older crater, Cyrillus or Catharina?

Seeing Lunar History

Look for the crater Kant. The meteoroid that created Kant was smaller than those that created Theophilus and Cyrillus–Kant is only 32 kilometers wide compared to the 100 kilometers of Theophilus. That meteoroid landed on a ridge just northwest of Theophilus, and under some light angles you can see the ridge wall rising up toward Kant's bright crater bowl. Kant lies in the "highlands" and even in a small telescope they actually *look* higher than the surrounding terrain.

But there's more here than just a crater hole on a hill. Between Kant and Catharina the ridge is hard to follow, but west of Catharina it becomes a noticeable scar curving south and east below Mare Nectaris. This scar, the Altai Scarp, is really the remains of the original impact basin that is now Mare Nectaris–certainly the largest and oldest "crater" in this group. North of the scarp, on what looks like the current shoreline of Mare Nectaris, you'll find two flooded craters, Beaumont and Fracastorius. Both have been filled with the lava flows that also filled the Nectaris impact basin. This is most obvious in Fracastorius, which looks like a small bay surrounded by short hills. Scientists believe that the Nectaris basin was created about 3.9 billion years ago, but filled with successive lava flows over the next 800 million years. Whatever created Fracastorius hit the moon after the Nectaris impact, but before the impact basin filled with lava.

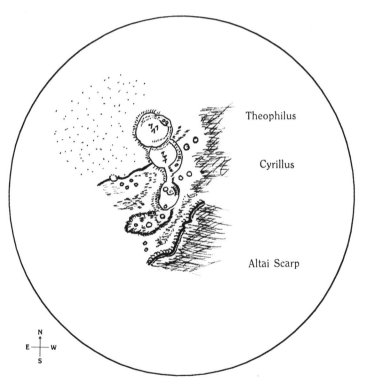

Fig. 13. Enlarged view of the Theophilus region,
80mm refractor @ 180x.

North is up and east is to the left in this drawing made when the terminator lay just west of the crater Kant. The clustered peaks and the inner terraces of Theophilus are *visible, as is the small crater Theophilus B on the northwest rim. A line of three peaks on the floor of Cyrillus runs parallel to the slumped western wall, on whose southwestern end lies the crater Cyrillus A. Northwest of Cyrillus are Ibn Rushd and Kant, and the larger crater west of the "channel" between Cyrillus and Catharina is Tacitus. Catharina's walls show only a hint of terracing on the northwest edge, where the vague circular form of the crater Catharina P meets Catharina's main wall. At this sunlight angle, the three craters curving south and east from Catharina appeared to lie on a ridge overlooking a featureless "valley." The Altai Scarp rises from the valley to the south. East along the shore of Mare Nectaris lies the flooded crater Beaumont, with some small bright drill hole craters nearby.*

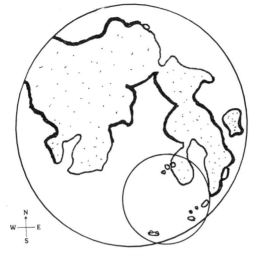

Expedition 3:
Set Off on Your Own to the Hommel Region

South of the Theophilus region there's a peculiar crater group that's worth exploring on the fifth day after a new moon. In the center of this group lies the oblong crater called Hommel. Hommel has been battered by large and small impacts over the ages. It is wider than Theophilus, but because it's on the southern end of the moon it is harder to see clearly all the details.

Ejecta blankets and the ruggedness and brightness of a crater's rim can be used to guess a crater's relative age. A bright ejecta blanket, or a well defined or bright rim, are signs that the crater is young. If a crater hole overlaps another crater hole, one of those craters is older than the other.

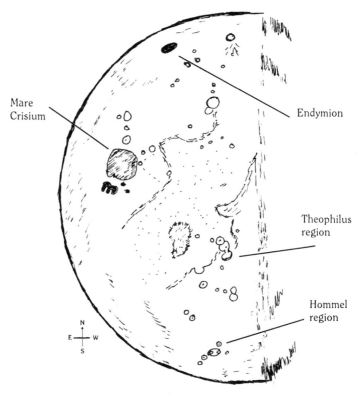

Mare Crisium

Endymion

Theophilus region

Hommel region

Fig. 14. 5-day-old moon

North is up and east to the left in this drawing of the 5-day-old moon, made with a 60mm refractor at 54x. The southeastern limb was lost in the glare, but to the north Mare Crisium and its surrounding craters were clearly visible. The Theophilus region, west and south of Crisium on the shore of Mare Nectaris, still shows some detail even at this relatively low power. The northernmost, dark crater is Endymion, and to the extreme south, just east of the terminator, lies the Hommel crater group.

Use these "rules of thumb" to analyze the Hommel region. Study this area under high magnification. Do Pitiscus and Vlacq have central peaks, or are those crater holes? The age difference between

Cyrillus and Catharina is obvious. Is there an obvious age difference between Vlacq and Rosenberger? Scan the rim and floor of Hommel itself. Which craters came first? Which came last?

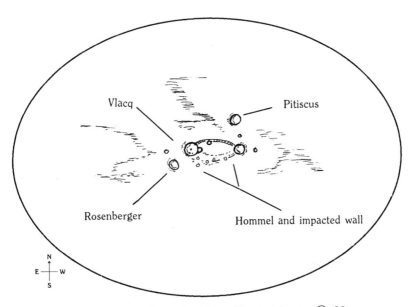

Fig. 15. The Hommel region, 60mm refractor @ 82x.

Expedition 4:
From Theophilus to Tranquillity Base
—The Apollo 11 Landing Site

On July 20, 1969, the *Eagle* separated from the *Columbia* on the far side of the moon as NASA and many of the rest of earth's inhabitants anxiously waited for radio contact to resume. The earlier Apollo 10 mission had tested the separation and docking procedures, but not the actual landing. Despite several on-board computer system alarms, astronauts Neil Armstrong and Edwin Aldrin descended toward the target zone on Mare Tranquillitatis. This was supposed to be a "safe" landing area. All the data from the Ranger, Surveyor, and Lunar Orbiter unmanned missions and the fly-over by Apollo 10 indicated a firm mare surface, a few rolling crater bowls, and few boulders.

Armstrong looked out the landing module window and saw boulders the "size of Volkswagens." Taking manual control, he steered out beyond the original landing zone. With only a few seconds of fuel left to steer the *Eagle*, Armstrong touched down on the southwest edge of Mare Tranquillitatis, at a spot now known as Tranquillity Base (Statio Tranquillitatis or Tranquillity Station on some maps).

Your trip should be a little less harrowing than theirs. First, following the example of the pre-Apollo missions, do some large-scale reconnaissance. Figure 16 shows all the Apollo landing sites. You'll visit all but the Apollo 16 site in these expeditions, starting with the Apollo 11 landing site (it's the easiest to find). Mare Tranquillitatis was originally chosen as the first landing site because it fit NASA's launch windows. It's a good site for our first expedition because

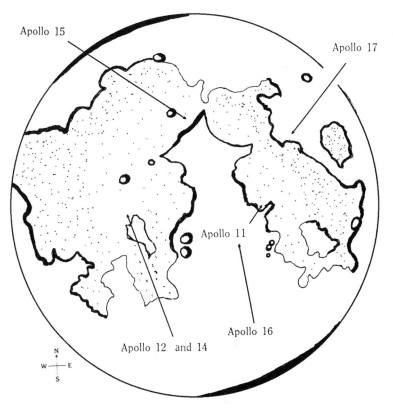

Fig. 16. The Apollo landing sites on the moon.

it fits our launch window—the new to first quarter moon.

Just to set your expectations, I should warn you that you won't see Armstrong's bootprints at this distance. Even the landing stage of the descent module is too small to see. With a small telescope, just finding the location can be quite a challenge. Bear in mind that in some ways you have a better view than Armstrong and Aldrin did. When they set foot on the mare they saw the lunar terrain from a perspective we will never share except in pictures. But even they saw the "magnificent desolation" of Tranquillity Base for only a few hours. You, on the other hand, can watch those mare plains for weeks at a time. One night you may see just the bright shoreline of Mare Tranquillitatis. The next night you may discover the sloping crater walls of Delambre and the ragged terrain to the south. Repeated observation is the next best thing to "being there."

How Do I Find It? Find the crater Theophilus. You explored this region in the last expedition. Theophilus juts out into the channel between Mare Nectaris to the south and Sinus Asperitatis to the north. (Sinus Asperitatis is the bay between Mare Tranquillitatis and Mare Nectaris.) Follow the coast of Sinus Asperitatis northward until you see a blunt peninsula. (There's another, pointed peninsula farther north—make sure you have the right one! You'll explore the pointy one later.) Use figure 17 to help guide you.

On the north side of this peninsula lies the Apollo 11 landing site. Figure 18 shows a close up of the area. The *Eagle* landed in the "curve" at the north side of the peninsula, midway between Sabine and Moltke.

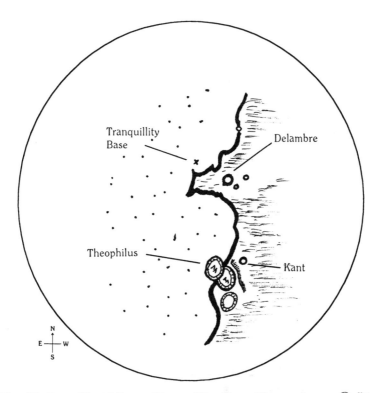

Fig. 17. From Theophilus to Tranquillity Base, 60mm refractor @ 54x.

Seeing Lunar History

It's 1964. You have been assigned to the astro-geological team of the United States Geological Survey. Your task: to find suitable landing sites for the Apollo missions. Mare Tranquillitatis was selected only after careful study by many pre-Apollo missions. But the original sites for pre-Apollo probing were selected based on visual observation. You have been assigned to "map" Mare Tranquillitatis for safe landing sites.

In 1964, lunar geologists assumed that the darker surfaces on the moon were younger terrain. Younger terrain meant that there were fewer craters. Fewer

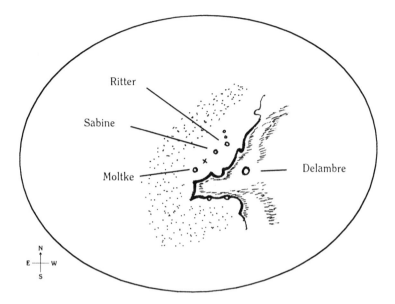

Fig. 18. Enlarged view of Tranquillity Base, 60mm refractor @ 82x.

craters meant less debris. Less debris meant "safer" landing zones. But sometimes across the dark, young terrain
there were light ray systems, in other words, ejecta from
crater impacts. Rays meant debris, and danger. So the
task was sometimes not to find the best spot, but instead to find the least dangerous spot to land.

Even in a small telescope you can "see" how difficult the task was. Observe Mare Tranquillitatis for
several nights in a row. Draw what you can see each
night, and note which features disappear and which
appear. Are there differences in the color of the mare
floor? Are there streaks that could be ray systems,
small hilly areas, or stubby ridges? Three years from
now, in 1967, NASA wants to soft land Surveyor 5 in
Mare Tranquillitatis. Surveyor's landing site should be
near a future possible Apollo landing site. Where
would you suggest NASA send the Surveyor 5 probe?

Before you leave Tranquillity Base, look for the terraced inner walls of the crater Delambre to the west of Tranquillity Base. Delambre lies in the "highlands"—in this case, the battered rims of the ancient Tranquillitatis and Nectaris basins. The terrain doesn't show much detail in a small telescope. This ancient surface has been eroded by impacts and covered by ejecta. Scan eastward back into Mare Tranquillitatis and look for the wrinkle ridges on the mare floor. These ridges are less bright than rays and only show up under the right sun angles. They aren't deposits of ejecta, they are distortions in the layers of lava that fill the mare.

N
W —┼— E
S

Expedition 5:
From Tranquillity to the Taurus Mountains
—The Apollo 17 Landing Site

On December 11, 1972, Astronauts Harrison Schmitt and Eugene Cernan piloted Apollo 17's lunar landing module *Challenger* into a valley southwest of the Taurus Mountains. NASA's intent was to collect geological data from diverse areas of the lunar surface. So previous Apollo missions had explored the maria, the southern highlands, and one "beach" between a mare and a mountain range. But many of the samples returned to earth by the Apollo astronauts were "contaminated" by ejecta from Mare Imbrium. Apollo 17's site, on the southeastern edge of Mare Serenitatis, was chosen in the hopes the crew would return samples from the pre-Imbrium moon and add a few more pieces to the puzzle of lunar history.

On this expedition, you will cruise the western shore of Mare Tranquillitatis, and then crater hop over to Apollo 17's landing site.

How Do I Find It? Use figure 19 to help find your way to the Apollo 17 landing site. This journey to the last Apollo landing site starts at the first landing site— Tranquillity Base. Head north along the western shores of Mare Tranquillitatis. The shore line is really the ancient rim of the Tranquillitatis impact basin. The original impact must have been tremendous. Look at how large Mare Tranquillitatis is even today after millions of years of erosion by meteors and lava flooding!

North of the Tranquillity Base peninsula there is another point of land where Mare Tranquillitatis meets Mare Serenitatis. This point of land is the

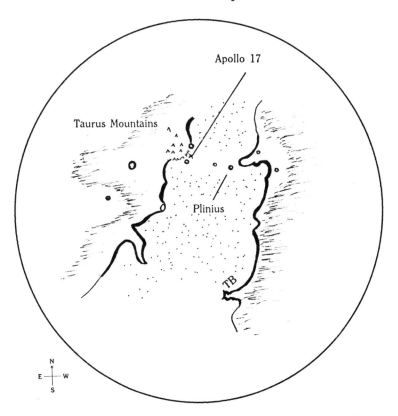

Fig. 19. View of the Apollo 11 and 17 landing sites, 60mm refractor @ 54x.

southeast edge of the Haemus Mountains. (Remember this place. You'll be back here later when you travel to Apollo 15's landing site.) At the east edge of this point of land, the Promontorium Archerusia, lies the bright crater Plinius, the launch pad for a short crater hop to join the *Challenger*.

Plinius has a high rim and a central peak, and it's easy to find in the channel between Serenitatis and Tranquillitatis. Just to the east is the crater Dawes, which is much smaller than Plinius. Draw an imaginary line from Plinius through Dawes to the shore on the other side of the channel. That line ends in the foothills of the Taurus Mountains.

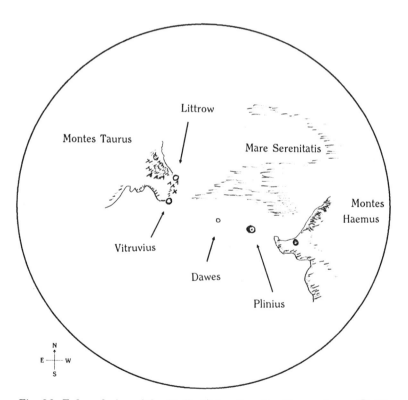

Fig. 20. Enlarged view of the Apollo 17 landing site, 60mm refractor @ 54x.

Now comes the hard part. The landing site lies between two craters, Littrow and Vitruvius. Both are small (about 30 kilometers across) and the floors of both have been flooded by lava. They can be hard to find because they don't have tall central peaks, and they don't have high rims like Theophilus. Littrow is the hardest to locate. Its south rim is damaged, and its west, north, and east rims look like ridges in the Taurus Mountains. Vitruvius is easier—it sticks out into Mare Tranquillitatis.

If you can find only Vitruvius, that's fine. Take three Vitruvius-sized hops directly north, and you've landed! The Apollo 17 lunar landing module touched down in a valley about two-thirds of the way between Littrow and Vitruvius.

The terrain here is more rugged compared to Tranquillity Base and it's difficult in a small telescope to see features clearly. But that's a significant difference! Think for a moment about where you just landed! This isn't the broad expanse of Tranquillity Base; it's a valley between mountains. In merely six flights, the Apollo program had progressed from large elliptical landing areas in expansive maria to pin point touchdowns in valleys a handful of kilometers wide. How good a driver were you your sixth time behind the wheel? And remember, those six flights had six different drivers!

Seeing Lunar History

What led NASA scientists to think that a landing near Serenitatis would provide clues about the pre-Imbrian moon? They had, of course, all the geological data from the Surveyor, Lunar Orbiter, and previous Apollo missions. But in some ways this data confirmed what some scientists had deduced just from visual observations.

Mare Imbrium lies hidden in the shadow west of the terminator during the new moon phase, so you'll have to wait until the first quarter to start exploring Imbrium. But if you retrace the paths you've taken in the last few expeditions, you can see the eastern half of this geological puzzle for yourself.

Your telescope is pointing at the Taurus Mountains and an outcrop of land (or "terra") between Mare Serenitatis, Mare Crisium, and Mare Tranquillitatis. Hop back across the channel and scan south along the west shore of Tranquillitatis. The terrain is visually similar to the Serenitatis shore: between the mare and the terra the color changes, but there's no large, well-defined rim as you might expect in a large impact basin. Moving farther south, you come to the Nectaris basin rim marked by the ridge beneath Kant and, still farther south, the Altai Scarp. But even this "rim" isn't very well defined and it doesn't run very far around Nectaris.

What held true for Theophilus and Cyrillus holds true of impact basins: the better defined the rim, the more likely it is a "young" rim. All three maria have poorly defined rims, and in fact these basins are about the same relative age (speaking in geological terms, of course). Nectaris is perhaps 3.9 billion years old, the Serenitatis basin was gouged out around 3.8 billion years ago, and both overlay Mare Tranquillitatis so that basin is older than both.

When the terminator moves westward and starts the first quarter phase, you'll have a chance to compare what you've seen here with a much younger formation on the west side of the puzzle. You'll also get to "see" why some of the other Apollo samples were "contaminated" with Imbrium ejecta. But while you wait for the terminator to move west, you can get in one more expedition on the eastern limb of the moon–the Mare Crisium area.

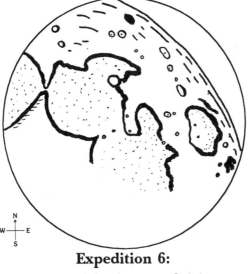

Expedition 6:
Hide and Seek Near Crisium

Up until now, your expeditions have been rather practical and scientific. You've learned how to crater hop and how to recognize age differences between features. You've followed in the footsteps of the Apollo 11 and 17 astronauts and you've observed some of the puzzle pieces that make up the geological history of the moon. It's time for a break. This expedition is just for fun.

The area around Mare Crisium contains many craters, a ray system, and two small seas. Some of these features are old and dilapidated, some are relatively new but small. What is visible depends on the moon phase and the angle of the sunlight, so observing this area several nights in a row is like playing hide and seek. Sometimes you see what you saw the night before, sometimes you don't. Some features like the small craters around Burckhardt and Cleomedes are best viewed early in the new moon phase. Other features like the Proclus ray system are visible through the full moon. But this area is worth scanning any-time you are out observing.

We have a lot of ground (sorry, regolith) to cover, so we'll break it into three tours.

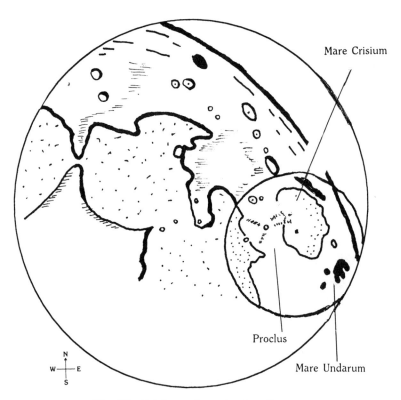

Fig. 21. Crisium–along the shoreline.

What to Look For Along the Shoreline: At the southeast edge of Mare Crisium, you'll see a dark "M" shape. This is Mare Undarum, the inland Sea of Waves. The crater Firmicus lies to its northwest, and Apollonius is to its west. Under the right sun angle, Firmicus and Apollonius appear to have dark blue floors and look like small circular seas. You know by now that this means they've been flooded with lava. To the northeast of Mare Undarum is another dark floored crater, Condorcet. It's more difficult to see because it is closer to the edge of the moon and its

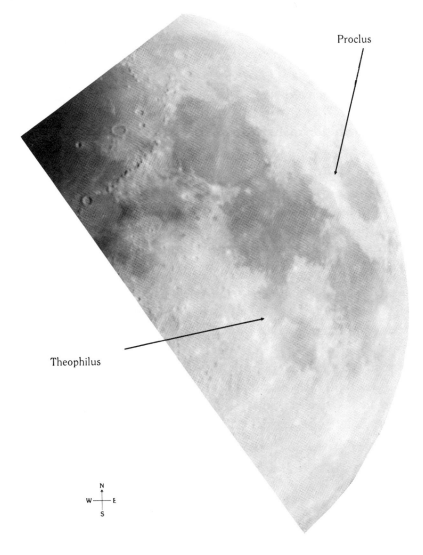

Proclus

Theophilus

N
W——E
S

Photo 2. Proclus crater.

*Oblong Mare Crisium and the rays of the crater Proclus
are the highlights on the easternmost moon in this photo taken
with a 5" Schmidt-Cassegrain telescope. The Taurus and
Haemus mountains are almost lost in the glare, but the crater
Plinius is still visible between Mare Serenitatis and Mare
Tranquillitatis. On the southwest shore of Tranquillitatis you
can see the peninsula where Apollo 11 landed, and the bright
central peak of Theophilus is hard to find but still visible on
the western shore of Sinus Asperitatis.*

walls are decayed. As the moon waxes, Condorcet is
the first of these three craters to be lost in the glare,
followed by Firmicus and Apollonius.

If Condorcet eludes you some night, you can always turn to the west shore of Mare Crisium. Here
you'll find the crater Proclus and its ray system. Two
prominent rays from Proclus lead eastward into Mare
Crisium, surrounding the crater Pierce. South of
Pierce there's the bright drill hole Picard.

Proclus, Pierce, and Picard are too small to show much
detail in a small telescope (all are less than 28 kilometers in
diameter). But if you look northward from Proclus you'll
see Macrobius. At 64 kilometers across, Macrobius is large
enough to show terraced inner walls and a central peak.
Directly east of Macrobius is the smaller and shallower crater Tisserand. Follow along the Crisium shoreline eastward
early in the new moon phase, and you may be able to see
Mare Anguis, the Serpent Sea—a blue squiggle behind the
rocky northeast shore of Crisium.

What to Look For from Cleomedes to Messala:
Cleomedes is the large crater north of Mare Crisium.
It has a central peak, but in smaller telescopes the
peak is easily missed because there's a small crater in-
side Cleomedes that "stands out" as if it were a peak.
Look for the two craters in the northwest wall (the

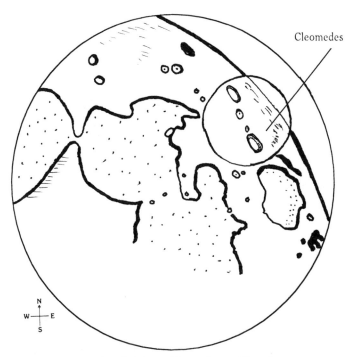

Fig. 22. Crisium–Cleomedes to Messala.

larger, westerly one is Tralles) and for the rille crossing northwest to southeast inside Cleomedes.

North of Cleomedes lies the trio of Burckhardt,
Geminus, and Messala (figure 23). Burckhardt impacted
on some existing craters, and its wall sometimes takes
on a different shape depending on the light angle.
Geminus has an elongated central ridge. So does Bernoulli to its east, but because Bernoulli is smaller its
ridge is hard to see in a 60mm refractor. North of Bernoulli is Bernoulli A, a bright but shallow crater leading
the way to Messala, a 124 kilometer-wide walled plain.
You'll need high magnification and a little luck early in

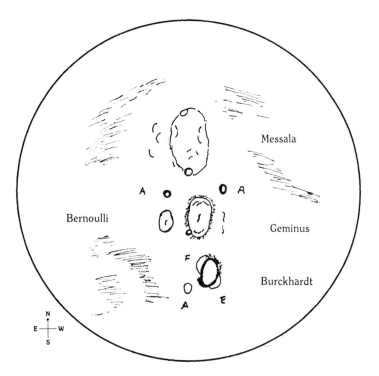

Fig. 23. Close-Up of Burckhardt, Geminus, and Messala, 60mm refractor @ 54x.

the new moon phase to see the ragged edges of the small and shallow craters inside and along its walls.

What to Look For on the North Slope: Scan west of Messala for the crater pair of Franklin to the south and Cepheus to the north. They are larger than Proclus, but they won't be as easy to spot as the moon waxes because their walls have slumped and they are rather shallow. Franklin has a central peak, and Cepheus is host to a small drill hole in its northeast wall.

Franklin and Cepheus point you toward Atlas and its western companion Hercules, both large enough to

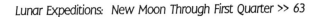

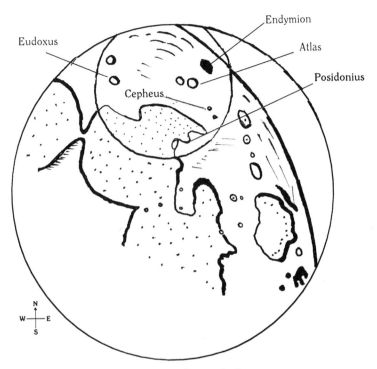

Fig. 24. Crisium–the north slope.

show some detail by the fourth day after a new moon. There are several rilles on the floor of Atlas, and the walls are noticeably terraced. The walls of Hercules also appear terraced, but instead of a central peak a small crater marks its floor. To their northeast lies the large, dark crater, Endymion.

Atlas and Hercules point you toward another crater pair, Eudoxus and Aristoteles, whose terraced and damaged walls are worth studying at high magnification. A line curving south and east from this pair brings you to what's left of Posidonius, with its internal rille system and a series of wrinkle ridges to its west.

The lunar surface west and north of Crisium is a mixture of craters, lakes, ridges, and plains. The terrain is flat, ragged, bright, dark—all depending on when you look at it. The curvature of the moon here makes it difficult to get your bearings, especially under high illumination. So don't be surprised if some night Franklin and Cepheus seem to have vanished and there are some unrecognizable ridges in their place. It's all part of the game of hide and seek played by viewing angle, light, and the notoriously fickle goddess of the moon....

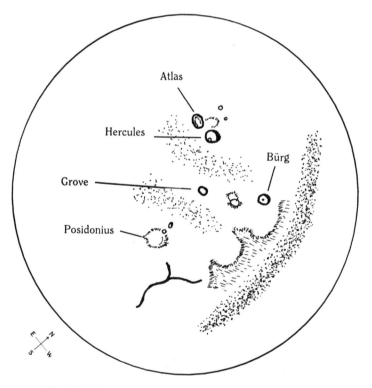

Fig. 25. Lacus Somniorum, 80mm refractor @ 120x.

North is to the upper right in this drawing made with the terminator crossing Mare Serenitatis west of the crater Hercules. The remains of Atlas E are visible between

Hercules and Atlas, and Hercules F is the small crater hole in the south wall of Hercules. Between Hercules and Posidonius to the south lie three craters: round Grove, the disintegrated craters Mason and Plana, and Bürg in the west with its barely visible central peak. Lacus Mortis, the Lake of Death, is the shaded area between Hercules and Bürg. Between Grove and Posidonius lies Lacus Somniorum, the Lake of Dreams. Daniell is the oblong crater north of Posidonius (no rilles in Posidonius were visible when this drawing was made), and the wrinkle ridge below Posidonius curves south and east into Mare Serenitatis.

N
W —|— E
S

*Photo 3. The southern lunar highlands. In the center of the
photograph lies the Ptolemaeus trio.*

Chapter 4
Lunar Expeditions:
First Quarter to the Full Moon

Expedition 7:
Hot Moon, Cold Moon—Ptolemaeus and Alphonsus

You looked at one crater trio on the five-day moon. Around the first quarter moon, another trio appears: Ptolemaeus, Alphonsus, and Albategnius. These three appear to lie on the northeast shore of Mare Nubium. An easy landmark in the first half of the first quarter moon, this crater group is more interesting for what you can't see than for what you can.

What to Look For: Figure 26 and photo 3 show where to locate the Ptolemaeus trio.

Ptolemaeus is the largest and northern-most crater of the three as shown in figure 27. There's no central peak, and the crater rim even in a small refractor is obviously badly decayed. You know by now that these are signs of an old crater. To the south lies Alphonsus, 120 kilometers wide. Its southern walls are better preserved than those of Ptolemaeus, and it has a central peak. It could well be the younger crater of the two. Just east of Alphonsus, the ring mountain

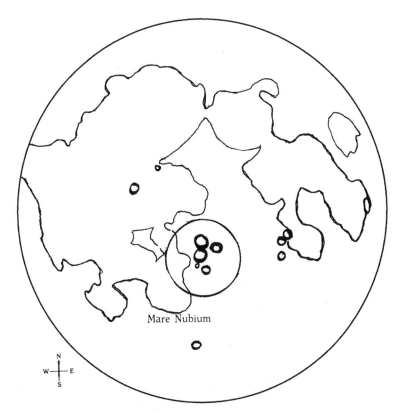

Fig. 26. The Ptolemaeus trio near Mare Nubium.

Albategnius feebly rises from the lunar surface. It surrounds the crater Klein, a more recent and better preserved impact hole. And in case you've forgotten what a young, "healthy" crater looks like, gaze to the south at Arzachel with its terraced inner walls and central peak.

But there are some intriguing differences between this grouping and the craters you saw in previous expeditions. Like Fracastorius, Ptolemaeus lies on a shore line. If it is an ancient crater, one might expect its rim to be damaged and its floor to be covered with lava just as Fracastorius has been damaged and

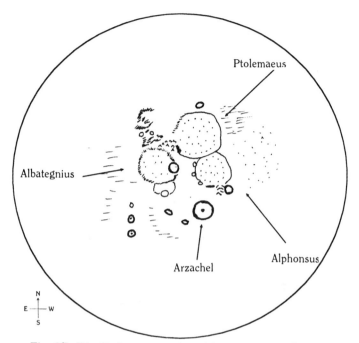

Fig. 27. The Ptolemaeus region, 80mm refractor @ 73x.

flooded by Mare Nectaris. But this huge 153 kilometer crater, though battered, still has a complete rim. Does this mean that Ptolemaeus, old as it may be, was gouged out of the mare after the mare basin had flooded? Possibly. But you know from Petavius that craters can be flooded and still have intact rims. Perhaps Ptolemaeus resembles Petavius over by Mare Fecunditatis more than it resembles Fracastorius over on Mare Nectaris.

Like Theophilus, Ptolemaeus has a very close companion. But the wall shared by Ptolemaeus and Alphonsus appears to be intact (although decayed), unlike the wall shared by Theophilus and Cyrillus. Theophilus and Cyrillus are both about 100 kilometers wide; Alphonsus is 120 kilometers wide. You'd expect the apparently larger Alphonsus impact to have broken down the rim of Ptolemaeus, as the Theophilus impact did to Cyrillus's wall. Does this mean Ptolemaeus is actually younger than Alphonsus? It doesn't *look* younger. Then why is its rim still intact? It could be that Ptolemaeus's walls were strong enough to withstand the Alphonsus impact. But there might be another explanation....

Seeing Lunar History

Is the moon "hot" or "cold"? Are there volcanic forces at work beneath its crust just as there are beneath the earth's crust? Or is the moon little more than a giant frozen rock, its internal magma fires having died eons ago?

According to the "hot moon" theory, popular before the Apollo missions, many craters on the moon were volcanic cones and vents, not impact holes, and some current features of impact craters were the result of volcanic activity after the impact. Ptolemaeus was a prime example. Under telescopic examination Ptolemaeus appeared roughly polygonal, a symmetry suggesting the crater and floor were created by volcanic eruptions–perhaps due to the Alphonsus impact. That would explain the old, intact rim. The Alphonsus impact cracked the lunar crust in the nearby, pre-existing crater hole, and whatever damage the impact inflicted on the old Ptolemaeus crater was immediately "repaired" by an eruption.

Alphonsus also showed evidence of volcanic activity. In 1958, the Soviet astronomer Kozyrev claimed he found carbon gas escaping from Alphonsus's peak, but his findings were never replicated.

In 1965, Ranger 8 photographed two other prime candidates for volcanic activity on its way to a crash landing in Mare Tranquillitatis. You saw these in the

Apollo 11 expedition–Sabine and Ritter (see figure 18). These visually young craters have no ejecta blankets and there are no secondary impacts nearby. Before Ranger 8, they were thought to be calderas. The Ranger 8 data was inconclusive.

A month after Ranger 8, Ranger 9 was sent to Alphonsus. Inside Alphonsus (beyond the grasp of small telescopes) lie several dark-haloed craters. Since they were inside Alphonsus, they had to be more recent than Alphonsus. Recent craters should be lighter than the surrounding terrain, not darker. Were these small volcanic vents?

Ranger 9's data was also inconclusive. The volcanic theory was neither proved nor disproved. Alphonsus remained a possible target of Lunar Orbiter or Apollo missions. When the Apollo 15 mission discovered dark halo craters in the Taurus-Littrow valley similar to the ones in Alphonsus, this valley became a prime landing site for Apollo 17. The Apollo 17 astronauts found that the dark halos were in fact partly composed of volcanic material, but this material had been excavated by impacts. There was no evidence for volcanic eruption, at least in Taurus-Littrow.

Alphonsus, on the other hand, remains a mystery even today. It was created before the Mare Imbrium impact, some 3.8 billion years ago. But the crater floor does not seem to be full of mare lava. And the photographic evidence has not conclusively proven that the holes in the crater floor are impact craters. Some observers claim to have seen Transient Lunar Phenomena (TLP) in Alphonsus—hazes or changes of color that might be volcanic outgassing. These sightings are probably just optical illusions. But no one will tell if you take another look at Alphonsus, just the same.

N
↑
W ——┼—— E
S

Expedition 8:
The Falcon Lands on the Fence

On July 30, 1971, the backup crew for Apollo 12 got their chance. Leaving Al Worden above in the Apollo 15 command module *Endeavour*, Dave Scott and Jim Irwin piloted the *Falcon* landing module past the towering peaks of the Apennines on their way to touchdown in Palus Putredinis, the Marsh of Decay. This mission would see extended "extra-vehicular activities" and the first use of the lunar land rover. And it would add a few more pieces to the puzzle of the moon's history and geology, as you'll see for yourself in this and subsequent expeditions.

The landing site lay between Mare Imbrium and Mare Serenitatis, one "young" and one "old" impact basin; perhaps the site would show stratified layers for the impacts and subsequent flooding of these basins. The Apollo 14 mission had returned Imbrium mare ejecta, but it had not collected samples of original lunar terra as expected. Perhaps these lay beneath the regolith in the valley below Hadley Mountain.

The apparently young craters Eratosthenes and Copernicus to the southwest might have spewed ejecta into this area that would help date those craters. The crater trio of Archimedes, Autolycus, and Aristillus was close enough to have left ejecta or secondary impact debris in the sampling area. Finally, there was the Apennine Bench, a light colored plain that appeared to pre-date the Imbrium basin lava flows. Was this perhaps part of the original lunar crust?

While the *Falcon* tottered on the side of the crater in which it had landed, NASA scientists waited for the pictures and samples that would help prove some theories about lunar geology and move others closer to extinction.

How Do I Find It? Figure 28 shows where the Falcon perched. Start from Tranquillity Base. Move north along the shore of Mare Tranquillitatis as if you were traveling to the Apollo 17 landing site again. When you get to the crater Plinius, don't hop the channel. Instead, go northwest along the Haemus Mountains. These visually "smooth and round" peaks rise only about 2,000 meters from the Serenitatis shore. Almost four billion years ago, they might have been the showpiece on the eastern half of the moon. But then an asteroid impact scoured out the Imbrium Basin and damaged this older impact basin rim.

At the northwest edge of Mare Serenitatis runs another channel. The south point of land is where the Haemus Mountains meet the Apennine Mountains. The north point of land is the end of the Caucasus Mountain range. Just west of this channel lie three craters: Archimedes, Autolycus, and Aristillus. South of these the Apennines slope down toward the crater Eratosthenes.

Here, on the "fence" between the Apennines and the Haemus ranges, between the Serenitatis and

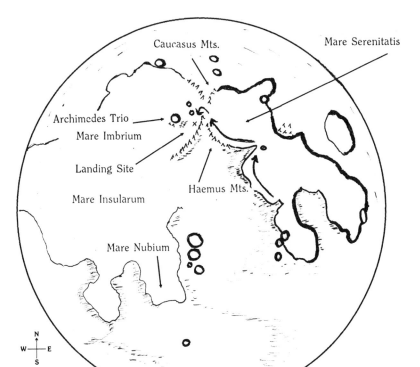

Fig. 28. The Apollo 15 landing site.

Imbrium basins, the Falcon perched for its geological excursion into the moon's middle ages. The close-up map (figure 29) shows approximately where Apollo 15 landed south of the crater Autolycus.

You'll explore the surrounding features in more detail later. For now, look for Conon, a 23 kilometer wide drill hole in the slopes of the Apennines.

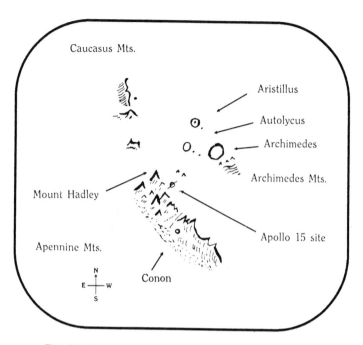

Fig. 29. Enlarged view of the Apollo 15 landing site,
80mm refractor @ 73x.

Seeing Lunar History

In the expedition to the Apollo 17 landing site, you traced the western shores of Tranquillitatis and Nectaris. You have probably already noticed that some of the terrain around the Apollo 15 landing site looks quite different.

If your observing schedule brings you to the Haemus-Apennine juncture early in the first quarter, you will notice that the Apennines really look like mountains. Whereas the Haemus range is squat and hilly, a look southward from Mons Hadley clearly shows peaks and a precipitous drop from peak to mare floor.

Better defined structures are younger structures, and indeed the Haemus range pre-dates the

Apennines on some estimates by millions of years. Since these mountain ranges are really impact basin rims, the Serenitatis basin is therefore older than the Imbrium basin. Try the same relative dating over on the eastern maria. Compare the Haemus range with the west shoreline of Mare Tranquillitatis. Which appears to be the older basin, Serenitatis or Tranquillitatis?

Eratosthenes

N
↑
W —|— E
|
S

Expedition 9:
Along the Imbrium Rim–South to Eratosthenes

Eratosthenes lies at the end of the Apennine range between Mare Imbrium and the southern seas, Cognitum, Insularum, and Nubium. By full moon, Eratosthenes is hidden beneath the bright ejecta from Copernicus, and the Apennines become a bright but relatively featureless ridge in a small telescope, so try to catch Eratosthenes on the eighth or ninth day after a new moon.

What to Look For: Start this expedition at the Apollo 15 landing site. Southwest of the landing site lies the Apennine Bench, your first stop.

Watch the Apennine Bench over several successive nights as the moon waxes from first quarter toward full. Visually, this small "plain" appears "corroded" yet rounded, suggestive of the Haemus range but obviously not so tall. It is surrounded by mare lava, suggesting it predates the lava flows. It is also light in color, suggesting it could be lunar terra. But just how did it get there?

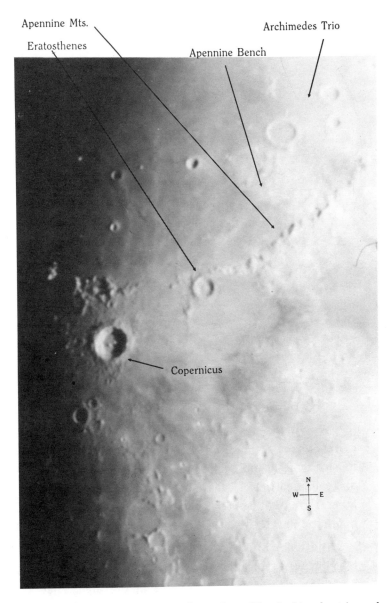

Apennine Mts.

Eratosthenes

Archimedes Trio

Apennine Bench

Copernicus

Photo 4. Dawn over the crater Copernicus. The Archimedes trio and the Apennine Bench are visible in the upper right corner.

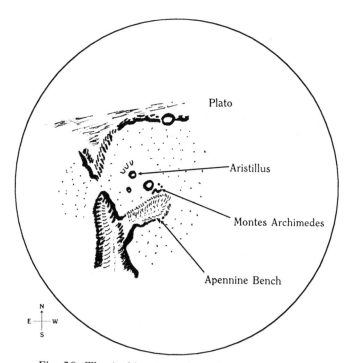

Fig. 30. *The Archimedes region, showing the rays near Aristillus and the Apennine Bench. 80mm refractor @ 72x, two days before a full moon.*

Seeing Lunar History

Was the Bench a volcanic upthrust through the mare? Was it a gob of impact melt from the Imbrium impact that slid up against the then new impact rim and solidified there? Was it an outcrop of the original lunar terra that somehow managed to survive the Imbrium impact?

The color and condition of the Bench indicates that if it were volcanic, whatever activity created it is far older than the lava eruptions in the area. That age doesn't rule out a volcanic origin. But volcanic activity that old wouldn't offer much support for a "hot moon" theory, either. The Apollo 15 crew did not sample Bench material directly. The samples from the landing site do, however, indicate that while there were two lava flows in Palus Putredinis about 3.3 billion years ago, the Bench was in fact created about 3.85 billion years ago, after the Imbrium impact. That means the Bench was in place before Aristillus was created and before the mare lava flooded the Imbrium basin and created Palus Putredinis. Sample analysis also suggests that the Bench may be the result of a volcanic eruption of primordial lunar magma–molten lunar terra, not mare lava.

The Bench is not the only vestige of the ancient moon that juts up through later lava deposits, as you'll see for yourself in the next expeditions.

East of the Bench, the portion of the Imbrium rim known as Montes Apenninus curves southwestward into Mare Imbrium. The Apennine crests, some 3,500 to 5,500 meters tall, are worth scanning nightly after the seventh day after a new moon. Peaks and passes appear and disappear under different light angles.

The Apennines end abruptly at the crater Eratosthenes. A young crater, it has a prominent ejecta blanket, terraced walls, and a central peak. The rays surrounding Eratosthenes are in fact rays from the crater Copernicus, which lies farther to the lunar west. In a full moon, Eratosthenes lies hidden beneath the brighter ejecta from Copernicus (which tells you something about the relative ages of these craters).

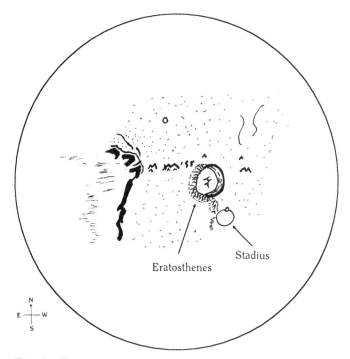

Fig. 31. Eratosthenes, Stadius, and the end of the Apennines. Ten-day-old moon, 80mm refractor @ 180x.

North is up and east to the left in this high magnification view of Eratosthenes. Several ravines in the Apennines are visible, shown as the white spaces between the dark clumps of peaks. Just north of the peaks, visible as a thin bright line, lies a small ridge "inside" Mare Imbrium. A small ridge runs south from Eratosthenes leading to the sunken crater rim of Stadius. Stadius B is the tiny drill hole in the north rim of Stadius. The wavy lines north of Stadius are two wrinkle ridges.

If you wish, you can launch north from Eratosthenes across Mare Imbrium to visit the crater Timocharis. Timocharis is the northwestern apex of a lop-sided triangle including Archimedes and Eratosthenes. All three are young craters, millions of years younger than the pre-Imbrium crater Alphonsus. But this trio shows that on the moon "young" doesn't mean "in perfect condition." Archimedes is flooded, indeed almost buried, by lava. The younger Eratosthenes is better preserved, but it too is buried, this time under ejecta from Copernicus. Timocharis escaped burial by debris or lava only to have its central peak destroyed by a subsequent impact.

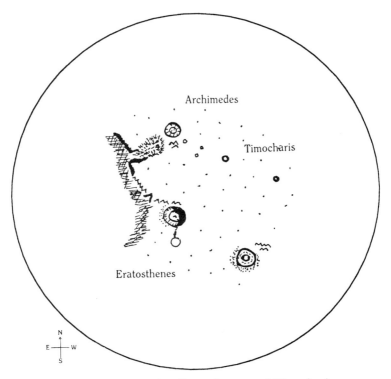

Fig. 32. Archimedes, Eratosthenes, and Timocharis. Nine-day-old moon, 80mm refractor @ 122x.

North is up and east is to the left in this wide field view of Eratosthenes. The ejecta blanket around the outer rim appears notched and pitted at this magnification, and two terraces are visible inside the crater itself. South of Eratosthenes the sunken crater Stadius appears to cling to a line of small ridges. South of Archimedes, the Montes Archimedes and the Apennine Bench jut from the surface of Mare Imbrium. Only Mons Wolff, at the end of the Apennine chain close to Eratosthenes, was distinguishable when this drawing was made.

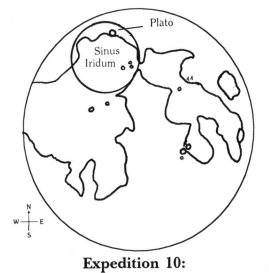

Expedition 10:
Across the Flood Plain—North to Sinus Iridum

As you've just seen, the southeastern rim of Imbrium has its showpiece craters and steep mountains. But the northern rim is not without attractions of its own. From the first quarter onward to the full moon, you can explore flooded craters, mountain ranges, unique ray systems, and even catch a glimpse of the Imbrium impact era lunar surface jutting up from the mare.

What to Look For: Start this expedition at the Apollo 15 landing site. Instead of heading directly south along the Apennines as you did in the previous expedition, head west to the crater trio Autolycus, Aristillus, and Archimedes.

At 39 kilometers in diameter, Autolycus is the smallest of the three craters. It marks the northwest end of Palus Putredinis. To the north lies the larger crater Aristillus, with its central peaks and ray system. This ray system is different from the one you saw by

Proclus: instead of straight lines, these rays appear "u" shaped.

The largest member of this trio is the crater Archimedes. It is some 80 kilometers in diameter, and the impact that created it spread ejecta across Palus Putredinis. But if you compare the floors of Aristillus and Archimedes, you'll notice that Archimedes has no central peak. Its floor was flooded by Imbrium basin lava. When young, Archimedes must have had an ejecta blanket and ray system like its younger sibling Copernicus to the south. But today, the most notable features are the Archimedes Mountains and some unrelated rays.

Seeing Lunar History

If you are thinking that Archimedes is the oldest crater of the trio, you're right. Its floor is flooded and it has no rays. On the other hand, the neighboring craters have either central peaks or ray systems or both.

By about 3 billion years ago, multiple lava floods in the Imbrium impact basin left Mare Imbrium and Palus Putredinis looking as we see them today. Autolycus and Aristillus lie on top of the lava, so those craters are younger than the final floods across the mare. Archimedes, on the other hand, is partially flooded. One could assume the Archimedes impact occurred on a lava surface, but before the last of the lava flows "topped off" the current mare.

If samples from the Apollo 15 mission are in fact ejecta from Aristillus or Autolycus, one of those craters is about 1.29 billion years old. Based on the visual evidence, Archimedes is therefore older than that. One might guess Cassini (or what's left of it) northeast of Aristillus, is older still.

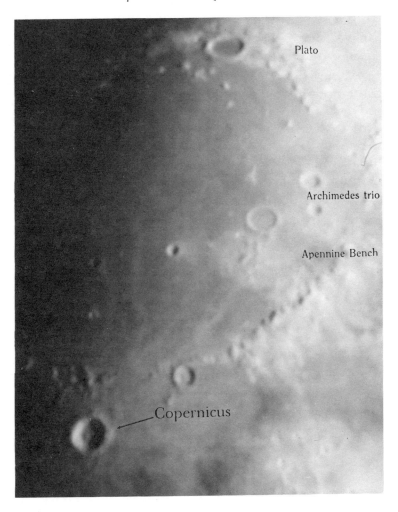

Photo 5. Plato, the Archimedes trio, the Apennines, and Copernicus.

Draw a line from Autolycus through Aristillus and you'll come to the Montes Alpes, the ragged northwest rim of Mare Imbrium. The Alps begin as a obvious rim formation but end as some scattered peaks in the "plains" east of the crater Plato between Mare Imbrium to the south and Mare Frigoris to the north. In the middle of the range, along the line drawn from

Autolycus through Aristillus, lies the 150 kilometer long Alpine Valley, looking like a trough or crease in the mountain range.

Where the Alps appear to flatten out, you'll find the crater Plato. Plato's rim (like that of Archimedes) did not collapse as the lava flooded the crater floor; unlike Archimedes, Plato shows no terracing or slumping. This makes Plato the perfect example of a lunar walled plain: 100 kilometers in diameter, the crater floor appears to be a smooth expanse of lava, and the rim appears sharp, unterraced, and intact.

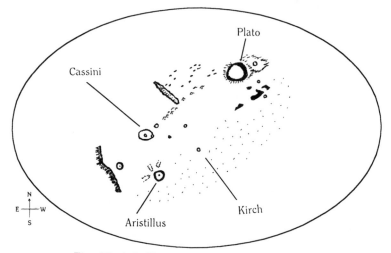

Fig. 33. Aristillus to Plato. Ten-day-old moon, 80mm refractor @ 120x.

In this panorama of the lunar surface, north is up and east is to the left. Some of the "u" shaped rays from Aristillus are visible against the mare background. Two of those rays point northwest toward the sunken crater Cassini, whose rim is barely visible above the mare surface. It is sometimes recognizable only because there are two bright drill holes inside it, Cassini A and B. To Cassini's west lies Mons Piton (the black dot) and farther out the two drill holes Piazzi Smyth (top) and Kirch (bottom). Individual peaks of the Montes Alpes stretch north and west from Cassini to the large crater Plato, broken in the center by the Alpine Valley. Plato A is the small crater to Plato's west. South of Plato, in Mare Imbrium, lie several small ranges: Mons Pico to the east, Montes Teneriffe in the center, and Montes Recti to the west.

Moving westward along the Imbrium rim, you come to Sinus Iridum, the Bay of Rainbows. Like Fracastorius, lava breached and covered a large part of Iridum's original crater wall, but Iridum is over twice as wide as Fracastorius and shows more detail. In a small telescope the rim looks like a finely etched setting for the dark bay. This etching is the Jura Mountains, the rim of the original Iridum impact. Like Eratosthenes, Sinus Iridum is an impact crater at the "edge" of the original impact basin. But unlike its younger cousin to the south, the Iridum impact occurred after the basin impact but before the basin flooded over. The Jura Mountains are all that is left of the original crater wall.

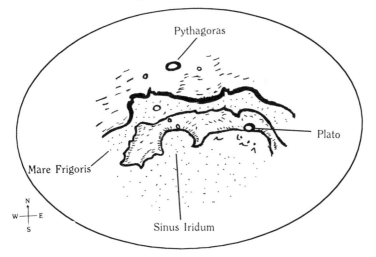

Fig. 34. Sinus Iridum and Plato. One day before the full moon, 4.25 RFT reflector @ 58x. .

In this drawing made through a reflector, north is up and east is to the right. The short focal length brings out the contrasts between the major features, but does not resolve smaller craters such as Plato A, which was visible in the previous drawing made with a long focal length refractor. Mare Frigoris, which marks the outer rim of the Imbrium basin, runs river-like between Pythagoras in the north and Iridum, the trailing edge of the Montes Alpes, and Plato to the south. Nestled in the Jura Mountains, which are the rim of Sinus Iridum, lies the crater Bianchini, and Foucault is visible to its northwest. Bright Harpalus shines from the middle of Mare Frigoris. Pythagoras is about 130 kilometers in diameter, larger than Plato, but because of its position on the moon's limb it is truncated and appears much smaller.

Seeing Lunar History

The Lunar Orbiter data on Mare Orientale on the lunar west limb show that this very young basin has three impact "rings." One would expect other basins to show this same structure. The Imbrium Basin rim (i.e., the Alps, the Apennines, and the north slope including Plato and Sinus Iridum) is probably the middle ring of the original impact structure. The western flows of Mare Frigoris, to the north of Plato and Iridum, may be the outer ring of the Imbrium impact. The inner rim has been almost destroyed by flooding.

Almost, but not completely. You may have noticed some bright, non-mare formations near the shoreline south and west of Plato. These are the Montes Recti and Montes Teneriffe, rising 1,800 and 2,400 meters above sea level respectively. (This is also the height of the Alps on the middle ring.) Like the Apennine Bench to the east, these two mountain ranges are vestiges of the lunar surface at the time of the Imbrium impact, although in this case the mountain ranges are part of the original excavated crust and not a volcanic eruption.

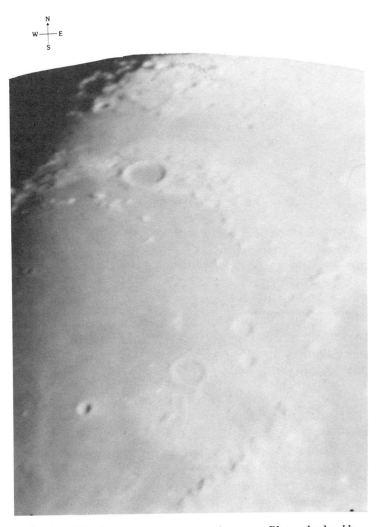

Photo 6. With the terminator west of the crater Plato, the knobby peaks of the Alps dot the Imbrium shore. The Alpine Valley cuts through the Alps to the east of Plato, and to the south and west the Montes Recti and Montes Teneriffe jut up from the Imbrium floor.

Fra Mauro

N
W — E
S

Expedition 11:
Fra Mauro and the Landing Sites of Apollo 12 & 14

Your last expedition to Apollo landing sites takes you to Mare Insularum, Mare Cognitum, and the Fra Mauro region. You probably recognize Fra Mauro as the target of the ill-fated Apollo 13 mission. The Apollo 12 and 14 missions, although successful, weren't exactly uneventful either: Apollo 12 was struck by lightening on liftoff, and the Apollo 14 lunar landing module suffered a temporary systems failure during the descent!

Like the Apollo 11 mission, Apollo 12 was sent to a flat, safe, mare landing site. But unlike Apollo 11, for which a "successful" landing might have been any safe landing, Apollo 12 had to prove that the lunar landing modules could be brought down on target. (You probably understand why this was so important now that you've seen the *Falcon* and *Challenger* sites!) On November 19, 1969, Charles Conrad and Alan Bean brought the *Intrepid* to rest about 160 meters from their landmark, the wreckage of Surveyor 3, southeast of the crater Lansberg.

The data from Apollo 11 had helped date some lunar formations but it had not settled the question of when and how the maria formed. Were they filled in at the same time, or were there different geological forces at work? Apollo 12 sampled a western mare, Mare Insularum, and the data showed that the lunar seas were created at different times. However cataclysmic the Imbrium impact might have been, it had not caused the only, or probably even the first, lava flood in lunar history.

Apollo 14 was sent to Fra Mauro to sample lunar terra, not mare lava; the hoped for prize was moon soil from beyond the dawn of time—or at least from before the lunar chronology dates that were known with any certainty in 1971.

What to Look For: Unlike many of the formations you've explored, Fra Mauro is "visually shallow" and it will take all your observation skills to wrench its secrets from it. Start with low magnification, because under some light angles and high magnification it is very easy to pass right over Fra Mauro without recognizing it. Figure 35 shows the location of the Apollo 12 and 14 landing sites.

From the Ptolemaeus-Alphonsus group, look west across the northern end of Mare Nubium for a small patch of bright terra shaped like a boot with a wide top cuff. Get the center of the boot's "upper" in your field of view, and then look *carefully*. The craters Fra Mauro, Bonpland, and Parry form a dilapidated triangle in the center of the formation. To the south lies small but "healthy" Tolansky and the barely visible rim of Guericke. To the north and west of Fra Mauro itself lies the "hummocky" section of the formation, so named for its low, ragged hills and poorly defined structures. Alan Shepard and Edgar Mitchell brought *Antares* down just north of the Fra Mauro rim. Under

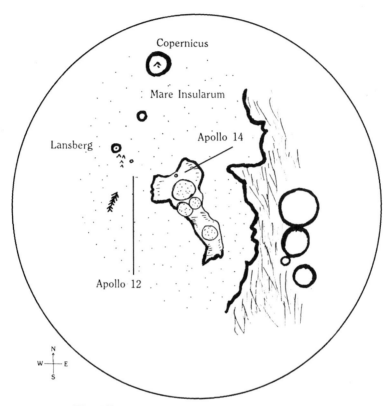

Fig. 35. Map of the region around Fra Mauro.

high magnification and glancing light, in a small telescope this area looks like a deeply etched mud flat. (The lunar rover wasn't introduced until Apollo 15. Shepard and Mitchell had to make do with a wheelbarrow. No, those are hummocks, not ruts.) Figure 36 shows this region as it appears through a refractor, and marks the Apollo 14 landing site.

Seeing Lunar History

Congratulations! You found the hummocky plains of the Fra Mauro Formation! Now just what *are* they?!?! You can tell by its light color that the Fra

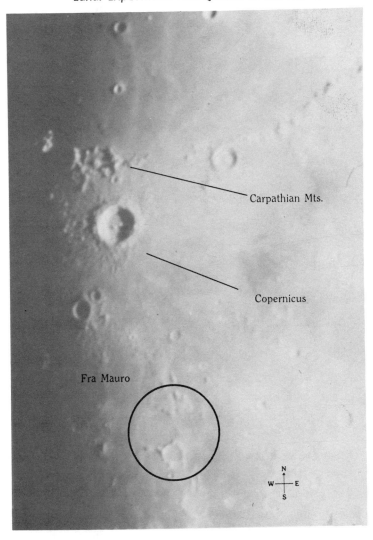

Carpathian Mts.

Copernicus

Fra Mauro

N
W — E
S

Photo 7. Easily overlooked in favor of Copernicus and the Carpathian Mountains (left center), Fra Mauro, Bonpland, and Perry (bottom center) are best identified not by their crater holes but by their shared wall.

Mauro Formation isn't a pure mare lava deposit. But the craters Fra Mauro, Bonpland, and Guericke look like they've been almost overflowed by lava. These are no round rimmed puddles like Archimedes or

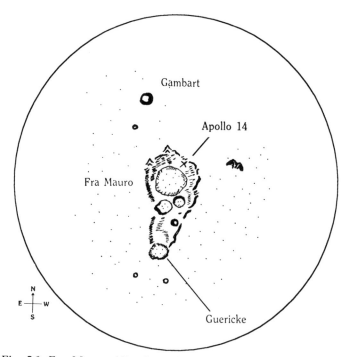

Fig. 36. Fra Mauro. Nine-day-old moon, 80mm refractor @ 122x.

Ptolemaeus; in fact they're not in much better shape than Cassini. Was Fra Mauro, like the Apennine Bench to the north, an outcrop of the ancient lunar terrain battered and then flooded over with Imbrium lava? Or was it a gob of Imbrium impact melt flung south from Imbrium?

Sorry, but you will have to decide for yourself. The Apollo 14 samples contained material older than the Imbrium impact, but it isn't certain how that material got there. It could have been older surface materials ejected by the Imbrium impact. Or it could have been an old Fra Mauro Formation substratum

**excavated by impacts on the formation and laid out
on top of Imbrium ejecta.**

From the Apollo 14 site, look west and north for
the crater Lansberg. Under glancing light and high
magnification, you may be able to detect a line of
"rocky" terrain slanting southeast from Lansberg.
Apollo 12 landed about two-thirds of the way between
Fra Mauro and Lansberg, at the end of this line.

Seeing Lunar History

South of Lansberg and west of Fra Mauro lies a
broken circle that at first sight appears to be a
breached, flooded crater. This is actually the Montes
Riphaeus, shown in figure 37. These mountains are
perhaps only a kilometer high. Like the Montes Teneriffe and Recti by Plato, they are probably the remains of an old impact basin. Montes Riphaeus lies
on the west edge of what today is known as Mare
Cognitum, an "impact" basin but not in the sense
you're used to. This area was named in 1964 after
Ranger 7 completed the first photographic mission to
the moon by smashing into the mare surface. If you
switch to low power and scan the shorelines of Mare
Nubium and Humorum, you'll see that Montes
Riphaeus is probably an outer rim of Nubium.

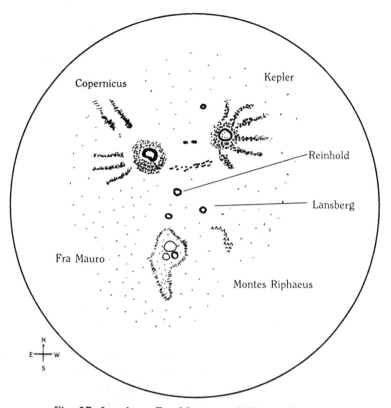

Fig. 37. Lansberg, Fra Mauro, and Montes Riphaeus.
Two days before a full moon, 80mm refractor @ 73x.

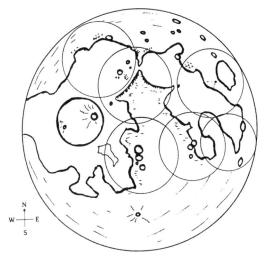

Expedition 12:
Copernicus—The Puzzle is Complete

In the early expeditions you traced the rims of the old impact basins Nectaris, Tranquillitatis, and Serenitatis. The last several expeditions showed you the younger, better defined Imbrium rim. Along the way, you learned how to relatively date features, and you have collected quite a few pieces of the puzzle of lunar history. One piece remains. We will look it over carefully, and then put the whole puzzle together.

The showpiece of the late first quarter moon is the crater Copernicus. At 93 kilometers in diameter it isn't the largest crater you can see, but its ejecta blanket and ray system more than make up for its size. It is the "classic" crater. In a small telescope, the basin appears deep and flat, surrounded by a bright, "sharp" rim. Inside, the crater rim is etched with terraces, at least two levels of which are visible under high magnification and steady skies. The central peaks are well preserved, appearing as a sharply defined "V."

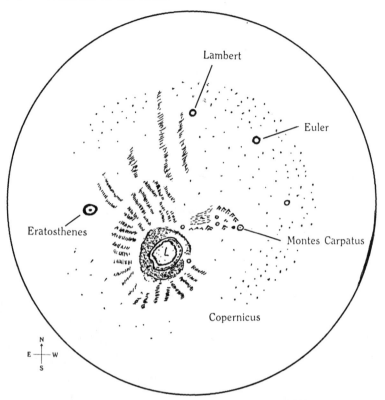

*Fig. 38. The crater Copernicus. Two days
before a full moon, 80mm refractor @ 122x.*

To the north of Copernicus lies the ragged terrain
of the Montes Carpatus rising a mere 1,000 to 2,000
meters above the Imbrium floor. Like their sibling
ranges the Alpes and Apennines to the north and
east, the Carpathians are the southern remains of the
original middle rim of the Imbrium impact basin.

North of Montes Carpatus lie the craters Euler
and Lambert. These craters form an arc with the craters Archimedes and Timocharis that points out into
the flatlands of Oceanus Procellarum, and only by co-
incidence seems to follow the curve of the Apennines
and Carpathians. West of Copernicus the crater

Kepler lies covered in rays from Copernicus but sporting its own bright, if shorter, ray system.

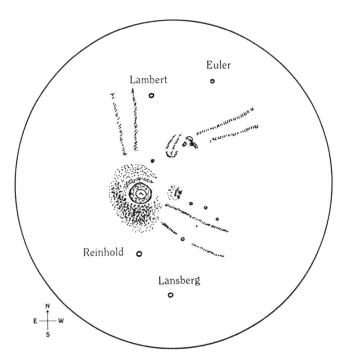

Fig. 39. Copernicus and Montes Carpatus.
Twelve-day-old moon, 60mm refractor @ 82x.

Under glancing light and high magnification even a small refractor will show details on lunar terrain around Copernicus. Surrounding Copernicus are patches of bright ejecta. On the outer walls of Copernicus the ejecta has a grainy appearance even in the 60mm refractor. Several rays stretch out across the mare surface where the ejecta thins. The rough terrain of the Montes Carpatus stretches north and west (right) of Copernicus, and several peaks were visible at the extreme west end. Directly below (south) of Copernicus is Reinhold, and to Reinhold's west lie a number of small craters just at the edge of visibility in the 60mm refractor.

The near full moon makes the lunar surface extremely bright. Theophilus and Ptolemaeus are just vague circles now; Petavius is barely visible. But you'll notice that Copernicus, Kepler, and even Lambert and Timocharis, stand out in spite of the glare. Brighter is younger, and indeed you are looking at features from the two most recent lunar eras.

Seeing Lunar History
Tables of Lunar Geological Eras and Basins

Lunar history is divided into five periods. While the historical dates for each period are still a subject of debate, the visual evidence for these periods is, as you've seen on these expeditions, sometimes quite striking.

Period	Duration	Notable Events
Pre-Nectarian	4.6 to 3.9 billion years ago	Heavy cratering and formation of some impact basins
Nectarian	3.92 to 3.85 billion years ago	The Nectaris Impact
Imbrium	3.85 to 3.2 billion years ago	The Imbrium Impact, multiple lava flows fill basins
Eratosthenian	3.2 to 1 billion years ago	Eratosthenes crater forms
Copernican	1 billion years ago to present	Copernicus crater forms

Based on the evidence compiled from telescopic observations and NASA's lunar missions, the mare basins can be dated with respect to the lunar eras (although sometimes the "dates" are only relative to each other). In the table below, those basins you've explored during your expeditions are listed by relative age, starting with the oldest basin in each period. Basins suffered successive major flooding up to perhaps 3.1 billion years ago.

Period	Basin Created
Pre-Nectarian	Procellarum Insularum Nubium Tranquillitatis Fecunditatis
Nectarian	Nectaris Humorum Crisium Serenitatis
Imbrium	Imbrium
Eratosthenian	
Copernican	

In previous expeditions you saw visual evidence for these estimations. The shoreline of Mare Tranquillitatis is less well defined than the shores of Serenitatis, and neither sport the rugged rims of Imbrium. Now as the full moon makes it difficult to examine the basins, you can turn your eye to the ray systems that highlight some of the younger craters on the moon. The rays of Copernicus and Kepler are splashed across Mare Insularum, Mare Imbrium, and Oceanus Procellarum. To the south Tycho's long spindly rays reach west across Mare Nubium and eastward as far as Mare Nectaris.

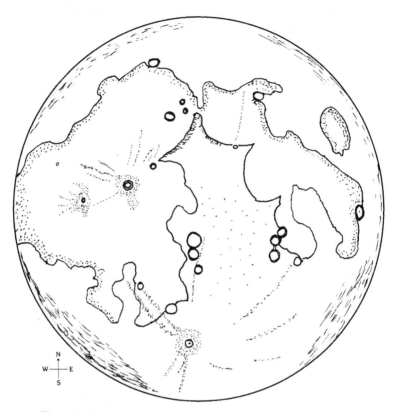

Fig. 40. Rays on the full moon, 4.25" RFT reflector @ 44x.

Seeing Lunar History
Tables of Craters by Lunar Eras

Bright ejecta blankets, ray systems, and the relative condition of a crater rim are just some of the evidence lunar geologists use to determine crater ages. Tycho, Copernicus, Kepler, and Proclus are some of the youngest craters you've seen. Is it possible to relatively date the other craters you've examined?

While exact dating is difficult without samples from the craters, the visual evidence can suggest when some craters were formed. Ptolemaeus and

Alphonsus are obviously older than Eratosthenes. Langrenus is younger than deteriorated Ptolemaeus, but older than bright drill holes like Conon. Archimedes lacks a ray system; Aristillus does not, so Archimedes is older than Aristillus. Based on photo-geological evidence as well as the Apollo samples, many of the craters you've explored can be placed into the five lunar periods. The tables below list the craters, their period, and the chapter where you examined them. Compare these relative age estimates with those you made on your expeditions. How do your findings compare with those of the experts?

The Pre-Nectarian Period

Crater	Expedition
Furnerius	Crater Chain on the Three Day Moon
Rosenberger	Set Off to the Hommel Region
Vlacq	Set Off to the Hommel Region
Messala	Hide and Seek near Crisium
Ptolemaeus	Hot Moon, Cold Moon
Fra Mauro, Bonpland, Parry	Fra Mauro

The Nectarian Period

Crater	Expedition
Pitiscus	Set Off to the Hommel Region
Tisserand	Hide and Seek near Crisium
Cleomedes	Hide and Seek near Crisium
Endymion	Hide and Seek near Crisium
Alphonsus, Albategnius	Hot Moon, Cold Moon

The Imbrium Period

Crater	Expedition
Petavius	Crater Chain on the Three Day Moon
Vitruvius	From Tranquillity to the Taurus Mountains
Macrobius	Hide and Seek near Crisium
Atlas	Hide and Seek near Crisium
Posidonius	Hide and Seek near Crisium
Arzachel	Hot Moon, Cold Moon
Archimedes	South to Eratosthenes
Cassini	North to Sinus Iridum
Plato, Iridum	North to Sinus Iridum
Lansberg, Gambart	Fra Mauro

The Eratosthenian Period

Crater	Expedition
Langrenus	Crater Chain on the Three Day Moon
Theophilus	Blankets and Rays
Plinius	From Tranquillity to the Taurus Mts.
Pierce, Picard	Hide and Seek near Crisium
Geminus	Hide and Seek near Crisium
Hercules	Hide and Seek near Crisium
Aristoteles	Hide and Seek near Crisium
Eratosthenes, Timocharis	South to Eratosthenes
Pythagoras	North to Sinus Iridum
Euler, Reinhold	Fra Mauro

The Copernican Period

Crater	Expedition
Proclus	Blankets and Rays
Eudoxus	Hide and Seek near Crisium
Dawes	From Tranquillity to the Taurus Mts.
Conon	The Falcon Lands on the Fence
Autolycus, Aristillus	South to Eratosthenes, North to Sinus Iridum
Kepler, Copernicus, Tycho	Fra Mauro, Copernicus

It's a pity that erosion and impacts have obliterated what must have been spectacular ray systems for such craters as Plato, Aristoteles, Ptolemaeus, or Cleomedes. The lunar observer can "look back in time" only so far as the features still exist. But the wealth of features on the moon will more than make up for the loss of a few spatterings of regolith over time. As you've worked through these expeditions you have probably seen lots of intriguing features along the way that weren't described in this book. So while you pack your telescope away for the night, let me suggest some other guides you might commission for your future expeditions.

WHAT TO DO AFTER YOU FINISH THIS BOOK

Start over again, of course! Remember, the scenery changes each night. Just because you've "been there," that doesn't mean you've seen all there is to see. Most handbooks for amateur astronomers have a section on the moon, and there are a few works devoted exclusively to lunar observation. Some especially good sources are listed below. (An asterisk indicates the book was a primary source in preparing *Welcome to the Moon.*)

Atlas of the Moon. Antonin Rükl. Kalmbach Books, 1990.*

Exploring the Moon Through Binoculars and Small Telescopes. Ernest H. Cherrington, Jr. Dover Publications, 1984.

A Field Guide to the Stars and Planets. J. M. Pasachoff & D. H. Menzel. Houghton Mifflin, 1992.

The Geologic History of the Moon. D. E. Wilhelms. U.S. Geological Survey Profession Paper 1348. U.S. Government Printing Office, 1987.*

An Introduction to the Study of the Moon. Zdenek Kopal. Gordon and Breach, 1966.

Lunar Sourcebook: A User's Guide to the Moon. Eds. G. Heiken, D. Vaniman, & B. French. Cambridge University Press, 1991.*

The Moon: An Observing Guide for Backyard Telescopes. M. T. Kitt. Kalmbach Books, 1992.*

Moon Shot. A. Shepard & D. Slayton. Turner Publishing, 1994.

Norton's 2000.0 Star Atlas and Reference Handbook. Ed. Ian Ridpath. John Wiley & Sons, 1991.

To a Rocky Moon: A Geologist's History of Lunar Exploration. D. E. Wilhelms. University of Arizona Press, 1993.*

GLOSSARY

Basin *The large depression left by the impact of a large meteoroid, usually filled with lava and identified as a sea or mare.*

Crater *A roughly circular depression usually caused by a meteor impact.*

Ejecta *The material ejected from the impact zone of a crater, containing scattered regolith, pulverized sub-regolith rock or mare deposit, and material melted by the heat at the impact point.*

Ejecta Blanket *The visual appearance of the ejecta that falls around the crater hole, usually containing older material than is found farther from the crater in the rays.*

Field of View *The apparent area visible through a telescope, which is larger or smaller based on the magnification and the apparent field of the eyepiece.*

Highlands *Lunar terrain raised above the level of the mare surface. May be the remains of an impact basin rim or lunar terra.*

Hummock *A low hill of ejected material with an irregular surface.*

Mare *A large, dark surface on the moon, usually filled with lava deposits; also called a sea.*

Rays *Material ejected by a crater impact and deposited in roughly linear formations across the lunar surface.*

Regolith *The dust, rock, and ejecta debris that makes up the lunar "top soil."*

Rille *Long cracks in the lunar surface, appearing as channels or as thin bright ridges depending on the sunlight angle.*

Rim *The uppermost edge of the crater hole; usually bright and well defined on young craters, often battered, slumped, or broken down on old craters.*

Ring Mountain *A large crater whose only prominent feature is a rim that rises substantially from the surrounding terrain.*

Scarp *A steep slope.*

Terra *A non-mare surface; also, the material that originally formed the lunar crust before huge portions of the crust were buried beneath lava flows.*

Terrace *A visual layer or strata inside a crater wall. In a small telescope these can be identified by a difference in brightness between the rim top and the crater floor.*

Walled Plain *A large crater whose floor seems level with the surrounding terrain.*

Wrinkle Ridge *Raised portions of the mare surface that look like wrinkles in small telescopes. These are of disputed origin.*

INDEX

MOON OBSERVATION REPORT

Date:	Time:	Moon Phase:

Low Power Magnification Sketch

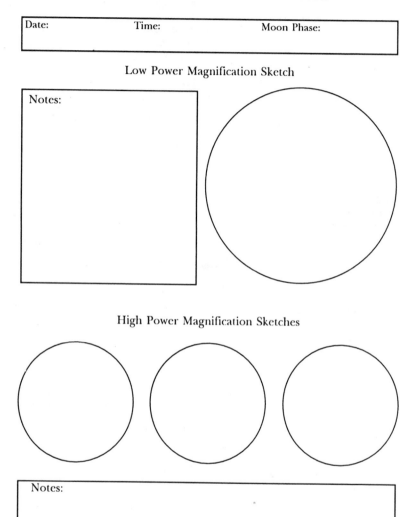

Notes:

High Power Magnification Sketches

Notes: